THE BIOLOGICAL ACTION OF PHYSICAL MEDICINE

CONTROLLING THE HUMAN BODY'S INFORMATION SYSTEM

THE BIOLOGICAL ACTION OF PHYSICAL MEDICINE

CONTROLLING THE HUMAN BODY'S INFORMATION SYSTEM

By

JAN ZBIGNIEW SZOPINSKI MD, PhD

Pain Clinic, Mayo Medical Centre, Constantia Kloof, Johannesburg, Republic of South Africa

AMSTERDAM • BOSTON • HEIDELBERG • LONDON
NEW YORK • OXFORD • PARIS • SAN DIEGO
SAN FRANCISCO • SINGAPORE • SYDNEY • TOKYO
Academic Press is an imprint of Elsevier

Academic Press is an imprint of Elsevier
The Boulevard, Langford Lane, Kidlington, Oxford, OX5 1GB, UK
225 Wyman Street, Waltham, MA 02451, USA

First published 2014

Copyright © 2014 Elsevier Inc. All rights reserved.

No part of this publication may be reproduced or transmitted in any form or by any means, electronic or mechanical, including photocopying, recording, or any information storage and retrieval system, without permission in writing from the publisher. Details on how to seek permission, further information about the Publisher's permissions policies and our arrangement with organizations such as the Copyright Clearance Center and the Copyright Licensing Agency, can be found at our website: www.elsevier.com/permissions

This book and the individual contributions contained in it are protected under copyright by the Publisher (other than as may be noted herein).

Notices
Knowledge and best practice in this field are constantly changing. As new research and experience broaden our understanding, changes in research methods, professional practices, or medical treatment may become necessary.

Practitioners and researchers must always rely on their own experience and knowledge in evaluating and using any information, methods, compounds, or experiments described herein. In using such information or methods they should be mindful of their own safety and the safety of others, including parties for whom they have a professional responsibility.

To the fullest extent of the law, neither the Publisher nor the authors, contributors, or editors, assume any liability for any injury and/or damage to persons or property as a matter of products liability, negligence or otherwise, or from any use or operation of any methods, products, instructions, or ideas contained in the material herein.

British Library Cataloguing in Publication Data
A catalogue record for this book is available from the British Library

Library of Congress Cataloguing in Publication Data
A catalogue record for this book is available from the Library of Congress

ISBN: 978-0-12-800038-0

For information on all Academic Press publications
visit our website at **store.elsevier.com**

Printed and bound in China

14 15 16 17   10 9 8 7 6 5 4 3 2 1

Working together
to grow libraries in
developing countries

www.elsevier.com • www.bookaid.org

Contents

The real voyage of discovery consists not in seeking new landscapes but in having new eyes. *Marcel Proust*

Preface vii
List of abbreviations ix

1. Introduction 1

1.1. Physical Medicine: General and Historical Background 3
1.2. Physical Medicine: Controversy About Definitions 5

2. Investigations of the Physiological and Morphological Foundations of Reflexive Physical Medicine 7

2.1. Anatomical Structure of the Skin 9
2.2. Known Functional Connections between Internal Organs and Skin 10
2.3. Thermographic Investigation of the Skin 10
2.4. Radioisotopic Investigation of Acupuncture Meridians 12
2.5. Influence of Organ Pathology on Bioelectrical Properties of the Skin 13
2.6. Investigations of the Histomorphological Structure of Acupuncture Points and Meridians 26
2.7. Discussion 27

3. Neurophysiological Foundations of Reflexive Physical Medicine 31

3.1. Introduction 33
3.2. Review of Relevant Data 34
3.3. Convergence Modulation Theory 40
3.4. Discussion 45
3.5. An Attempt to Visualize Organ Projection Areas 46

4. Organ Electrodermal Diagnostics 53

4.1. Implementation of Optimal Measuring Parameters for an OED Device 55
4.2. Localization of Particular Organ Projection Areas 56
4.3. Clinical Assessment Of OED Accuracy 63
4.4. Discussion 70

5. Reflexive Physical Therapies 73

5.1. General Remarks 75
5.2. Thermotherapy 77
5.3. Hydrotherapy 80
5.4. Ultrasound Therapy 80
5.5. Phototherapy 82
5.6. Electrotherapy 88
5.7. Magnetotherapy 106
5.8. Reflexive Mechanical Stimulation 109

6. Final Considerations 223

References 227
Index 231

Preface

ABOUT THIS BOOK

This evidence-based manual of modern physical medicine is written primarily for physiatrists (specialists in physical medicine) and also for all the other medical doctors interested in this specific field of medicine; according to the American Association of Academic Physiatrists, a medical doctor must be trained in physical medicine in order to direct a comprehensive rehabilitation team of health professionals that may include physiotherapists, occupational therapists, psychologists, speech therapists, social workers, and others. It has to be emphasized, however, that in the United States and some other countries physical medicine is combined with medical rehabilitation into one specialty, so the term "physical medicine" has become a synonym in those locations for rehabilitation. In European countries, by contrast, physical medicine and medical rehabilitation are two separate medical specialties; physical medicine is sometimes combined with balneology and climatology (in Germany and Poland), but not with medical rehabilitation. This book is devoted to "genuine" physical medicine (see Section 1.2 in Chapter 1), not to medical rehabilitation.

Nevertheless, physical medicine can be useful across the spectrum of medical disciplines. Neurologists should benefit from this book; they are trained to make proper diagnoses in their scope of expertise, but on many occasions there is not much they can offer to their patients even when it comes to the treatment of such common conditions as persistent migraine, neuralgia, neuropathy, or severe back pain. Anesthesiologists, especially those who specialize in chronic pain management, should find this book useful as well. Physicians can extend their effectiveness, because not all medical problems can be helped sufficiently by using pharmaceuticals alone. Rheumatologists, ophthalmologists, ENT specialists, and other professionals will learn that many problems in their respective fields of expertise can be managed more effectively with physical therapies. Even general practitioners will find this book helpful in their family practices; more efficient GP practices will, in turn, ease the burden of hospitals.

This book should also be of interest and help to physical therapists and those physiotherapists who are engaged with physical medicine. It will definitely be of benefit to all of the so-called alternative or complementary health practitioners who often use certain methods of physical medicine without sufficient scientific background. Neurophysiologists, biophysicists, and biomedical engineers might also find the content of this publication worthwhile.

A great deal of controversy and sometimes prejudice still surrounds physical medicine in general and so-called reflexive therapies in particular. Some of these therapies are even regarded as "alternative" or "complementary" medicine, leading valuable research articles in the field to be automatically rejected by renowned medical scientific journals. On the other hand, the public in general fully recognizes the value of physical medicine, and medical practices specializing in this field are in demand. Due to the shortage of such professional services in many countries, however, in almost every shopping mall various electrical, mechanical, thermal, and other therapeutic and pseudotherapeutic devices are offered, sometimes at high prices. Also, various homegrown healers all over the world widely advertise their suspicious diagnostic and therapeutic services. Naturally, all these unproven devices and services give physical medicine a bad name.

The main reason for the existing situation seems to be the fact that most physical medicine methods have gained acceptance, but their mode of action is still not fully clear. There have been various attempts to explain all the particular types of "Western" physical therapies: thermotherapy, hydrotherapy, phototherapy, ultrasoundtherapy, electrotherapy, and magnetic field therapy. However, all of these theories, which can be found in existing manuals of physical therapy and in instruction manuals of various therapeutic devices, in fact produce more questions than answers. On the other hand, traditional Far Eastern beliefs concerning the so-called bioenergetic therapies are unacceptable from a scientific point of view. There is no unified, convincing, and scientifically acceptable physiological theory to date that can be universal for all methods of physical medicine and is in accordance with the contemporary state of biological and medical knowledge.

In order to address this complicated situation, since the early 1980s our successive research teams conducted an independent comprehensive research program aimed at bringing some scientific order to still chaotic medical discipline. This evidence-based approach resulted in the specific structure of this book. In order to substantiate claims, the first part reviews relevant research and in this way creates scientific foundations for practical aspects, both diagnostic and therapeutic, that are discussed in the second part. The idea is that readers who familiarize themselves with the first part will subsequently find all the right answers to all the potential questions of the field on their own. For practical reasons, the less important research is summarized briefly, whereas crucial research is described comprehensively with all the details so that it can be easily checked and confirmed by other researchers.

It is worth mentioning that the whole program of research conducted over the last 30 years and described in this book was funded from the limited private resources and performed "after hours" by groups of enthusiasts who saw great potential within the program. There are many wonderful people who contributed to this research in various ways and who deserve the highest possible gratitude. However, because it would be an endless list, allow me to mention here at least those who made the most significant impact on this project. First, I want to thank Prof. Gerard Jonderko, former head of the 4th Department of Internal Medicine at the Silesian Medical University (Katowice, Poland), who not only taught me medicine but also infected me with a passion for scientific research. I also want to thank the late Prof. Zbigniew Garnuszewski, a great European expert in classical acupuncture, who shared his outstanding knowledge with other "Western" doctors. I am most grateful to the late Prof. Tadeusz Mika and Prof. Gerard Straburzynski, my postgraduate teachers of physical medicine. I thank Prof. Stefan Wegrzyn, former head of the Institute of Informatics at the Silesian Technical University (Gliwice, Poland), and his team of outstanding electronic and computer engineers for building the first automatic prototype of the organ electrodermal diagnostics (OED) machine. I am grateful to the Polish Ministry of Health for organizing and sponsoring my scientific trips to the leading acupuncture centers in China, Korea, and Mongolia as well as the National Institute of Reflexotherapy in Moscow. I want to express my gratitude to the South African Government for financial assistance with the preparation, patenting, and certification processes of the OED device's final model. Finally, my highest appreciation goes to my former research partner, electronic engineer Tadeusz Sierak, with whom I set the foundations for the current research program, and to my current research partner, Georg Philip Lochner, whose groundbreaking research (described in his M.Eng. dissertation) led to the comprehensive molecular explanation of the skin reversible electrical breakthrough effect and rectification phenomenon.

Jan Zbigniew Szopinski

List of Abbreviations

AC	alternating current
AP	acupuncture point
Cit.	cited in
CNS	central nervous system
DC	direct current
ELF-MF	extremely low frequency – magnetic fields therapy
IR	infrared radiation
LLLT	low level laser therapy
LRP	low resistance point
OED	organ electrodermal diagnostics
OPA	organ projection area
PP	pressure point
PSSO	pain syndrome of spinal origin
TENS	transcutaneous electrical nerve stimulation
TP	trigger point
UV	ultraviolet radiation

CHAPTER 1

Introduction

OUTLINE

1.1 Physical Medicine: General and Historical Background — 3

1.2 Physical Medicine: Controversy About Definitions — 5

1.1 PHYSICAL MEDICINE: GENERAL AND HISTORICAL BACKGROUND

Things of nature, including the human body, are a mixture of chemical reactions and physical phenomena. In practice, this means that medicine should explore equally both ways, i.e. chemical one and physical one, for diagnostic penetration and therapeutic intervention. On the diagnostic side, proportions are still preserved: physical methods, including basic physical examination, X-ray, magnetic resonance imaging (MRI), ultrasound and nuclear diagnostics, electrocardiogram (ECG), electroencephalogram (EEG), electromyogram (EMG), etc., are in use along with various chemical laboratory tests, but when it comes to therapy, a strong trend is observed in contemporary medicine to treat everything in the chemical way. This is mainly due to spectacular pharmacological achievements in the 1940s and 1950s: since then, medical students have been trained predominantly in biochemistry and very little in biophysics. This has resulted in a situation in which medical doctors in general do not speak the same language as engineers, despite the fantastic achievements of technical sciences in more recent years. Currently most medical practitioners seek to see pathology exclusively in chemical aspects, ignoring the fact that, for example, the nervous system's mode of action can be explained much better by physical sciences than by chemical reactions. A gap has been created in contemporary medicine between what we have now and what we could achieve, taking into account vast progress in technique and physical sciences.

Historically, physical therapies are as old as humankind. At a very early stage, people discovered the therapeutic effects of heat, cold, solar radiation (heliotherapy), and water application (hydrotherapy). Descriptions of early heliotherapy and hydrotherapy can be found in the writings of two famous ancient doctors: Hippocrates of Cos (460–380 B.C.) and Asclepiades of Bitinia (120–56 B.C.). In China, cold baths were used for fevers as early as 180 B.C. Ancient Romans created foundations for spa treatments, and some of their famous spas are still in use today. In their famous "terms," in addition to certain forms of hydrotherapy, therapeutic massage was also practiced. An auto-massage of painful body parts comes almost as a reflex after any injury; but in ancient Greece and Rome, massage was in use not only as a therapy but also for "sports medicine" purposes. There is also evidence that ancient Greeks applied electric fishes to painful areas of their bodies, creating in this way the first form of electrotherapy. In the Middle East, "Turkish steam baths" were introduced, combining elements of both thermotherapy and mechanical stimulation.

Physical therapies played a particularly positive role when early "pharmacotherapy" often brought more harm than good (e.g., an overuse of mercury and blood-letting in medieval Europe). In the 16th century, Paracelsus (real name: T.B. von Hohenheim, 1493–1541) displayed a great deal of interest in magnetotherapy. In 1776, the American doctor E. Perkins built a magnetic field-based device for pain relief. Also in the 18th century, the first electrophysiological experiments by Luigi Galvani (1737–1798) and the creation of the first electric cell by Alessandro Volta (1745–1827) prompted immediate experiments with electrotherapy. In 1831, M. Faraday (1791–1861) discovered the phenomenon of the electromagnetic induction and in this way made possible wide use of "faradic current" in electrotherapy. At the end of the 19th century, J.A. d'Arsonval (1851–1940) and N. Tesla (1856–1943) discovered high-frequency currents, which are of utmost importance in contemporary electrotherapy. In the 20th century, high-frequency electrical and magnetic fields were introduced to medicine (shortwave and microwave diathermy).

Discoveries of infrared (F.W. Herschl, 1800) and ultraviolet radiation (J. Ritter and W.H. Wollaston, 1801) are the milestones in phototherapy. In 1895, Danish doctor N.R. Finsen (1860–1904) used ultraviolet emitted by a self-constructed lamp for treatment of skin TB. Albert Einstein's (1879–1955) quantum theory created a basis for the development of the laser technique. Low-power lasers were introduced to physical medicine in the 1960s.

In 1880, the Curie brothers discovered a piezoelectric phenomenon, leading to the possibility of the use of ultrasounds in medicine. Since 1951, ultrasoundtherapy has become one of the most popular methods of physical medicine.

Observations made by the Silesian farmer V. Priessnitz (1799–1851), followed by the scientific research of W. Winternitz (1835–1917), created foundations for modern hydrotherapy. German priest S. Kneipp (1821–1897) went even further, recommending various forms of hydrotherapy combined with a hygienic lifestyle, diet, and exercises for almost any disease, including syphilis and other infectious conditions.

The early 20th century still belonged to physical medicine, with the proliferation of famous Swiss and other European spas—for example, Baden-Baden, Karlovy Vary, and many others—in which various diseases including arthritis, peptic ulcers, coronary heart disease, asthma, and even TB were treated with physical therapies. Hot wrappings were widely applied for poliomyelitis.

However, the most amazing development in physical medicine came from the Far East. It seems that ancient Chinese doctors discovered that certain, sometimes remote, skin spots become tender in the case of a disease of a particular internal organ. This tenderness disappears after the organ is cured. The doctors must have presumed that a connection exists between these spots and the related organs, so they started to stimulate these skin areas therapeutically not only with deep point massage (Tien-An/Shiatsu) but also with needles (acupuncture) and heat (moxibustion). Observing the good clinical effects of these procedures, they developed a fundamental principle: "all the tender skin areas should be stimulated." Over time, precise acupuncture maps were created, and the points corresponding to the same organs were connected with the artificial lines called meridians. Having no physiological and very little anatomical knowledge, Eastern practitioners adapted their general dualistic cosmic theory of the antagonistic vital energies "yang" and "yin," which apparently circulate in the human body along the meridians, in order to explain how acupuncture works.

More recently, the French doctor P.M.F. Nogier, working in North Africa, observed locals treating their domestic animals, including horses and camels, by cauterizing particular zones on the animals' ears. He examined human ear auricles and concluded that a particular area would become tender when a related internal organ is diseased. In this way, he created the first maps of auricular organ projection areas and originated the concept of the auricular homunculus with the shape and position generally similar to the early fetus. He also employed this discovery for therapeutic purposes by inserting small acupuncture needles or applying laser radiation at these points.

Interestingly, traditional Southern African healers who have never heard of acupuncture perform therapeutic procedures of so-called *scarifications* by making superficial incisions at particular skin areas close to a diseased organ. For example, small cuts are made on the chest to alleviate bronchial asthma and around the stomach or the knee to reduce persistent pain in these areas.

There must be something special about acupuncture; after six thousand years, it is still in use for a number of conditions. Hundreds of research articles published in various peer-review scientific journals indicate the high clinical effectiveness of this oldest system of physical medicine. Among the international journals dedicated specifically to medical acupuncture are: the *Journal of Traditional Chinese Medicine, Medical Acupuncture,* and *Deutsche Zeitschrift fur Akupunktur.*

These days, "Western" and "Far Eastern" methods of physical medicine work together; specific skin areas are stimulated not only mechanically with needles, stitches, cupping, or point massage, but also chemically (injections, plasters, or creams), with ultrasound, laser, and other forms of phototherapy, with cold/heat (thermotherapy), and with magnetic fields and various forms of electrostimulation—from the transcutaneous electrical nerve stimulation (TENS) to the shortwave/microwave diathermy. Most of these therapies have gained acceptance from mainstream medicine, but their mode of action is still not fully elucidated.

1.2 PHYSICAL MEDICINE: CONTROVERSY ABOUT DEFINITIONS

Physical medicine means the medical management of diseases and disorders using various forms of physical energy. Many medical disciplines use physical methods in their daily practice. For instance, surgeons in general use electrocoagulation, laser knives, and cryosurgery; neurosurgeons in particular use radiofrequency rhizotomy; psychiatrists use electro-convulsive therapy; cardiologists use defibrillation; and dermatologists use lasers. Medical rehabilitation uses electrostimulation for paralyzed muscles, and chiropractors use manual manipulation for spinal problems. Certain physical treatments have even become official medical specialties on their own: for example, oncological radiotherapy. However, *there is a group of so-called reflexive therapies, which use various forms of physical energy to stimulate and control the body's own self-defense mechanisms and systems. Because of their very specific mode of action, these therapies emerge as a separate discipline of clinical medicine: reflexive physical medicine.* Typical reflexive therapies include: thermotherapy (heat or cryostimulation), phototherapy (infrared, ultraviolet, or laser), ultrasoundtherapy, electrotherapy (direct or via electromagnetic energy), magnetotherapy and mechanical nerve stimulation (acupuncture, reflexive massage, cupping, or high-pressure hydrotherapy). Chemical stimulation of the skin's nervous receptors by the use of various injections (even bee stings), plasters, compresses, and creams also belongs to the same category, because respective chemical substances are utilized in this case as nervous stimuli and not as medications of their own. We must always remember that, for example, in the case of electrotherapy it is not the electrode and in the case of acupuncture it is not the needle which cure the problem; these are just tools to stimulate and control the body's own powerful self-defense mechanisms and systems. After all, the human body is, in fact, the best possible "pharmaceutical factory," which under nonpathological circumstances is able to synthesize any needed substance.

Certain traditional indications for reflexive physical medicine—for example, ischemic heart disease, hypertension, peptic ulcer disease, diarrhea, urinary tract infection, impotence, etc.—are no longer indications for reflexive physical medicine due to tremendous progress in contemporary pharmacology. However, there is a wide range of pathological conditions for which reflexive therapies should be still the first choice. For instance, these therapies are particularly well suited to treating severe back pains, even after multiple failed spinal surgeries. They are known to be very successful when it comes to treatment of persistent headaches, neuralgias (intercostal, trigeminal, or postherpetic), neuropathies (for example, diabetic), phantom pains, reflex sympathetic dystrophy, Reynaud syndrome, chronic rhinitis/sinusitis, Meniere syndrome, tinnitus, hearing loss due to acoustic nerve damage, Bell's palsy, glaucoma, macular degeneration, atonic urinary bladder. They can be visibly effective as a supportive treatment of Parkinsonism. Reflexive physical medicine is traditionally used for various kinds of arthritis, ankylosing spondylitis, acute gout, and sports injuries; it can be very useful in the case of chronic respiratory tract diseases. Those more radical reflexive therapies, especially electroacupuncture and cryostimulation (−70 to −160 degrees Celsius), can be successfully used even in acute problems: for example, status asthmaticus, atonic uterus, renal colic, esophagospasm, severe migraine, or postoperative pains. Electroacupuncture can be also by far the most successful treatment for nicotine addiction.

Of course, it is good in practice that physiatrists (specialists of physical medicine) know various physical therapies useful in rehabilitation—for example, traction, spinal manipulation ("manual medicine"/chiropractice), electrostimulation of muscles, or radiofrequency rhizotomy. However, these treatments belong to the medical rehabilitation specialty (or even neurosurgery), and nerve blockades and intra-articular injections are respectively the domains of anaesthesiology and orthopedic surgery. *What really constitutes the difference between the specialty of physical medicine and other medical disciplines, which also use certain physical treatments, are reflexive therapies. Therefore the term "physical medicine" should be used with regard to reflexive physical medicine rather than, for example, medical rehabilitation.*

It is also important to distinguish between the terms physical medicine and physiotherapy. Physical medicine comes from physics, in contrary to the term physiotherapy, which originates from the Greek "physis"—nature—and should be reserved mainly for "physiological" therapies such as exercises (kinesitherapy, biokinetics, water exercises, etc.). Whenever any therapeutic devices stimulating the sensory nervous system are used, it should be called physical medicine, with the respective medical specialists involved regarded as physiatrists and other respective therapists regarded as physical therapists. For prophylactic, therapeutic, and rehabilitative purposes, balneotherapy uses certain natural materials, for example, mineral waters, peloids (resembling mud), and therapeutic gases. Treatments are usually applied in the form of the bath (sometimes inhalation), combining both the elements of physical medicine (heat or hydrotherapy) and the pharmacotherapy, with the leading therapeutic factor being the chemical composition of specific substances absorbed through the skin. If mineral water is used internally in specific dosages and at specific times, it is called krenotherapy. Climatotherapy includes natural elements of physical medicine: for example, thermotherapy and heliotherapy (solar radiation).

CHAPTER 2

Investigations of the Physiological and Morphological Foundations of Reflexive Physical Medicine

OUTLINE

2.1 Anatomical Structure of the Skin   9

2.2 Known Functional Connections Between Internal Organs and Skin   10

2.3 Thermographic Investigation of the Skin   10

2.4 Radioisotopic Investigation of Acupuncture Meridians   12

2.5 Influence of Organ Pathology on Bioelectrical Properties of the Skin   13
    2.5.1 History of Investigations   13
    2.5.2 Investigation of Skin Electrical Potentials   16
    2.5.3 Investigation of Skin Electrical Resistance with a Standard Ohmmeter   17

    2.5.4 Influence of Organ Pathology on the Electrical Impedance and Resistance of Organ Projection Areas of the Skin   18
        Patients and Methods   19
        Results   21
    2.5.5 Influence of Organ Pathology on Bioelectrical Properties of Acupuncture Meridians   24

2.6 Investigations of the Histomorphological Structure of Acupuncture Points and Meridians   26

2.7 Discussion   27

2.1 ANATOMICAL STRUCTURE OF THE SKIN

Various forms of physical energy, used in physical medicine to stimulate and regulate the body's own powerful self-defense mechanisms and systems, are usually applied to the skin. Therefore it is important to properly understand the anatomical structure of the skin (see Figure 2.1).

The skin consists of distinct principal layers: the epidermis, the dermis, and the subcutaneous layer. The dermis and subcutaneous layer contain the vascular and nervous components of the skin as well as the sweat glands, sweat ducts, and hair follicles. The epidermis contains no blood vessels and is entirely dependent on the underlying dermis for nutrient delivery and waste disposal via diffusion through the dermoepidermal junction. The epidermis is a stratified squamous epithelium consisting primarily of keratinocytes in progressive stages of differentiation from deeper to more superficial layers. As keratinocytes divide and differentiate, they move from this deeper layer to the more superficial layers. Once they reach the stratum corneum, they are fully differentiated keratinocytes devoid of nuclei and are subsequently shed in the process of epidermal turnover. Cells of the stratum corneum are the largest and most abundant of the epidermis. The stratum corneum consists of approximately 15–20 layers of corneocytes (flattened cells filled with keratin, originally keratinocytes) that are surrounded by lipid lamellae, consisting typically of five stacked lipid bilayers (see Figure 2.2).

From the physical medicine point of view, it is important that the stratum corneum has been shown to be by far the most important electrical current barrier (94). Tape stripping of the skin has shown that when the stratum corneum is removed, the rest of the epidermis and dermis can be modeled as a resistance of only 500 ohms. The stratum corneum and the appendages, which traverse it, are therefore primarily responsible for the electrical properties of the skin.

Skin appendages such as sweat glands and hair follicles traverse the stratum corneum. Sweat ducts are approximately 10 µm in diameter and account for approximately 0.1% of the skin surface. Sweat ducts are presumed to act as shunt pathways for electrical current through the skin.

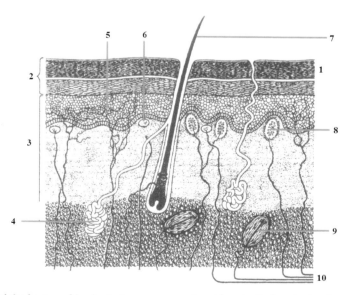

FIGURE 2.1 Typical structure of the human skin. 1: stratum corneum, 2: epidermis, 3: dermis, 4: sebaceous gland, 5: free nerve ending, 6: Krause-type corpuscle, 7: hair, 8: Meissner-type corpuscle, 9: Vater-Pacini-type corpuscle, 10: sensory nerve.

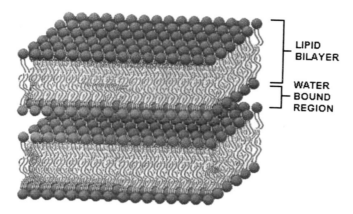

FIGURE 2.2 Lipid bilayer.

2.2 KNOWN FUNCTIONAL CONNECTIONS BETWEEN INTERNAL ORGANS AND SKIN

Various relationships between internal organs and the skin are known. Skin sensitivity, temperature, hydration, and color may be changed by internal organ pathology. Also, skin electrical parameters are influenced by many internal diseases. The impact of endocrinological function and autonomic innervation of particular dermatomes on the skin electrical resistance is well known. The correlation between psychological status and skin electrical resistance, known as the psychogalvanic reaction, is utilized in the so-called polygraph test ("lie detector"); mental stress, caused, for example, by lying, almost immediately increases perspiration, resulting in significantly diminished skin resistance, especially on the fingertips.

On the other hand, a number of reflexive therapies are known in contemporary medicine: therapies in which the skin is stimulated in order to obtain specific medical benefits not related to the skin itself. The oldest reflexive therapies historically were acupuncture (needle stimulation), pressopuncture/acupressure (point massage), and "moxa" (heat application). At a later stage, cupping was added. Classical acupuncture relatively recently developed auricular medicine (stimulation of the auricular skin areas) as well as craniotherapy (scalp acupuncture). Classical massage, also relatively recently, included specific foot and hand skin zones ("reflexology"). Contemporary physical medicine stimulates skin nervous receptors not only mechanically with needles, stitches, or point massage, but also chemically (by injections, plasters, or creams), with ultrasounds, low-power lasers, and other forms of phototherapy, with cold/heat (thermotherapy), with magnetic fields, and with various forms of electrostimulation, from transcutaneous electrical nerve stimulation (TENS) to shortwave/microwave diathermy.

2.3 THERMOGRAPHIC INVESTIGATION OF THE SKIN

Thermography is used in medicine in order to assess the temperature of various areas of the body. Higher temperatures can be caused by increased metabolism, for example, inflammatory pathologies or muscle spasms; they can also indicate areas of intensive blood supply. Lower temperatures are usually displayed by the areas of poor local blood circulation. Thermovision cameras show areas of higher temperature in brighter colors, whereas colder areas are marked by darker colors.

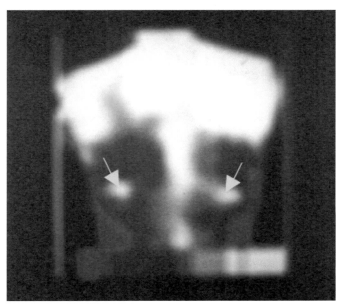

FIGURE 2.3 Thermogram of an upper-back pain sufferer (posterior aspect). Upper-back muscles and spine display higher temperatures (bright regions). Two bright trigger points are visible below the scapulas at the location of the classical Yishe acupuncture points.

The flow of any known kind of energy always produces a thermal effect, and changes in one kind of energy always cause changes in other kinds of energy. Presuming that hypothetical acupuncture meridians are channels for some kind of energy ("vital energy"), one could expect to see them with thermography as lines of higher temperature. Also, some authors (cit. 20, 48, 61, 78) believe that acupuncture points (APs) are skin areas of increased metabolism; that is, these skin spots should also be seen with a sensitive thermovision camera.

Therefore, tests were performed (70, 76) on a group of inpatients with various clinical diagnoses and on a group of clinically healthy volunteers at the 4th Department of Internal Medicine, Silesian Medical University, Poland, using an AGA-Thermovision Model THV thermographic camera (accuracy 0.1 degrees Celsius):

- Skin temperature along so-called meridians was assessed in each case from various angles.
- For diseased persons, small acupuncture needles were inserted in five classical APs located on meridians related to diseased organs, according to classical acupuncture rules (always including so-called stimulating or sedative points, depending on the kind of pathology) and three skin spots located outside of meridians. For healthy volunteers, the needles were inserted in five classical APs located on the so-called Large Intestine Meridian (always including the famous point LI 4, known as "Hegu") and three skin spots located outside of meridians. The temperatures of these skin areas were monitored for 15 minutes each.
- Inserted needles have been stimulated in various configurations with DC and AC of various parameters (below the pain threshold) under continuous thermographic monitoring.

During the temperature assessment along the meridians, no areas of increased temperature (which could correspond to meridians) were found. However, so-called trigger points ("ashi" points)—that is, skin areas of increased sensitivity to digital pressure, usually located on top of muscle spasms—displayed significantly higher temperatures within areas of 20mm in diameter (Figures 2.3 and 2.4). Trigger points (TPs), which still belong to classical APs, are not painful on

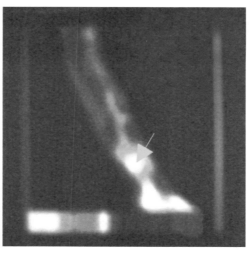

FIGURE 2.4 Thermogram of the left forearm (internal aspect) of a coronary heart disease patient. A bright trigger point is visible above the wrist at the location of the acupuncture point Neiguan.

their own; they are simply very tender, so the patients generally were not aware of their existence prior to examination.

Insertion of the needle increased the temperature around the needle within the range of square millimeters by 0.23 ± 0.1 degrees Celsius. Insertion of needles in other skin spots did not affect the temperature of skin areas already punctured. Electrical stimulation did not cause any temperature changes.

Our thermographic investigations of the skin confirmed neither the existence of acupuncture meridians as energetic channels nor an increased temperature of all classical APs. However, thermography can localize TPs: the increased temperature of these specific skin areas could be caused by an underlying muscle spasm.

2.4 RADIOISOTOPIC INVESTIGATION OF ACUPUNCTURE MERIDIANS

Some authors (cit. 20, 48, 61, 78) believed that acupuncture meridians could be channels for so-called cosmic radiation, that is, electromagnetic energy emitted by the sun. The frequency of this radiation is similar to that of radioactive isotopes. North Korean researcher Kim Bong-Han (2) claimed in the 1960s that by using a scintigraphic technique, he was able to prove the existence of another system of the body: the Kyungrak System, which in certain ways could be identified as a system of meridians. However, the images he presented were in fact those of blood/lymphatic vessels.

Our investigations (70, 76) have been done on a group of inpatients with various clinical diagnoses and a group of healthy volunteers at the 4th Department of Internal Medicine, Silesian Medical University, Poland. In each case, two measuring probes SSU-70-2 with a scintillator NaJT1040x25 (background 750 imp/s, exploitation 105000 imp/s) of the measuring device ZM-703-M3 (manufactured by Polon) were placed along various meridians (Figure 2.5). For diseased persons, the meridians of choice were those corresponding to diseased organs, according to classical acupuncture rules. For healthy volunteers, the meridian of choice was the Stomach Meridian. The

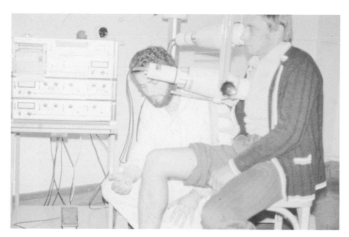

FIGURE 2.5   Radioisotope injection into an acupuncture meridian, monitored with sensitive detectors of gamma radiation.

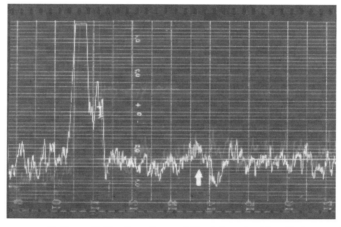

FIGURE 2.6   Time record of the gamma-radiation activity along the meridian injected with radioisotope. The time of injection is marked with an arrow; the "peak" visible prior to injection was caused by placing the radioisotope-filled syringe in proximity to the measuring probe.

radioisotope 113m In was injected successively into three chosen APs located on the investigated meridian (500 µCu per each point). For diseased persons, the stimulating or sedative points of respective meridians have been always included, depending on the pathology. For healthy volunteers, the S 36 point (known as "Zusanli") was always included.

If meridians were really conductors of cosmic/radioactive energy, then one could expect an immediate increase in the radiation activity along the respective meridian. However, the radioisotope injections did not show such an effect in any of the previously described cases (Figure 2.6).

2.5 INFLUENCE OF ORGAN PATHOLOGY ON BIOELECTRICAL PROPERTIES OF THE SKIN

2.5.1 History of Investigations

Many authors have investigated the effect of organ pathology on the electrical parameters of the corresponding skin areas. Croon and Overhof measured the electrical resistance and capacitance in strictly estimated skin areas (cit. 22, 59). In this way, they tried to localize the organic pathologies and secondary functional disorders and also active primary focal infections. Wolkewitz evaluated electrical resistance, capacitance, and temperature along certain lines on the patient's skin, in different body areas (cit. 23, 61). In this way, he tried to localize the primary focal infection. Many authors supported the Galvanopalpation Test established by Gehlen and Standel as a complementary test in localization and evaluation of inflammation of certain internal organs (13, 31, cit. 6, 23, 61). The skin area over the examined organ is stimulated with a weak direct current (at the individual perception threshold). If the organ is inflamed, flaring of the stimulated skin area is observed. Shimmel, meanwhile, created the Segmentary Electrogram (cit. 30, 33, 61). He examined the impedance of certain body regions by using eight six-centimeter-diameter electrodes. On the basis of these measurements, he tried to localize diseased organs and determine the type of pathology.

A number of diagnostic methods were created using specific bioelectrical properties of APs. In 1929, Dumie built a device which was supposed to localize APs on the basis of diminished electrical resistance in comparison to surrounding areas (cit. 61, 78). Also, Niboyet conducted investigations of various bioelectrical phenomena connected with acupuncture (41, cit. 61). He suggested that both APs and acupuncture meridians (lines connecting APs related to the same internal organs) display diminished electrical resistance and higher potential. Kreczmer, Dumitrescu, and Nicolau also indicated the difference between the resistance, capacitance, and potential of APs and neighboring skin areas (cit. 61, 78). Podshibiakin, meanwhile, stated that APs display tenderness, increased temperature, and electrical potential (cit. 48, 61, 78). They are also characterized by increased absorption of oxygen. "Biologically active points," which include APs, display potentials 2–3 mV higher than surrounding skin, and these potentials rapidly increase with acute inflammatory diseases. In the case of chronic diseases, these potentials go down; after healing, the potentials normalize. The author connected these changes with the function of the nervous system. According to the author, diagnostic results could be affected by external factors such as ionization of the air, the solar phase, and the psychological status of the patient. The number of biologically active points with diminished electrical resistance was much greater than the number of classical APs.

Meerzon and Kotlar suggested that the electrical impedance of APs differs from that of other skin points (cit. 48, 61). Rosenblatt based his electropuncture diagnostic method on the characterization of APs' impedance across a wide range of measuring parameters (53). Bratu, Prodescu, and Georgescu created a diagnostic method based on the measurements of electric resistance and impedance of particular APs (5). By means of this method, they attempted to identify disturbed "meridians." According to these authors, pathology was supposed to diminish the electrical resistance of APs. Similar changes in the electrical resistance of particular APs, connected to the character and location of pathological processes, were observed by Nagiyeva, Dunayevskaya, Kassil, and Ivanov (cit. 48, 61).

Some of these authors emphasized that changes in the electrical resistance of related APs occurred earlier than clinical symptoms of the disease, which indicated the possibility of early diagnostics and monitoring of the pathological process dynamics. Nikiforov and coworkers described the phenomenon of asymmetry in electrical conductivity characteristics of APs (cit. 51, 78). In certain APs, greater conductivity was observed for negative polarization of the measuring electrode and, in others, for positive polarization of the same electrode. Increased resistance for a positively polarized measuring electrode was supposed to indicate pathology with an excessive amount of hypothetic "bioenergy," whereas the reverse phenomenon accompanied a shortage of this energy.

"EAV" by Voll (88, cit. 33, 48, 78) with modifications "Vegatest" (cit. 30) and "B.E.S.T" (17) as well as "Ryodoraku" after Nakatani (25, cit. 48, 78) with modification "CITO" (48, 78) all belong to the most popular methods of electropuncture diagnostics that use corporal APs. "EAV" is based on the hypothetic interaction between currents circulating in particular "meridians" and currents circulating in a measuring circuit. In order to identify the location of the "bioenergy disturbance," the author recommends measurement of electrical resistance of 172 selected points using a 1.5–2 V measuring voltage. Nakatani utilized the hypothetic decrease in the electrical resistance of APs related to diseased organs. His diagnostic method, Ryodoraku, also evaluates "meridians" using round, 1 cm^2 measuring electrodes. Nakatani believed that the bioelectrical changes observed in APs are due to the influence of the sympathetic nervous system.

There is an increased interest in auricular electropuncture since the publication of the map of auricular organ projection areas (OPAs) by Nogier (42). On this map, particular zones correspond to only one internal organ. According to Nogier, auricular OPAs related to diseased organs display reduced electrical resistance. Experimental localization of auricular OPAs corresponding to particular organs was done on animals by Niboyet, Kwirchishwili, and Portnov (48). However, there are significant differences in maps of auricular OPAs prepared by different authors (15, 20, 34, 35, 42, 48, 77, 78). Balaban and Rozenfeld suggested that auricular OPAs exist due to anastomoses between the vagus nerve and other nerves supplying the ear auricle (cit. 48). Similar hypotheses were proposed by Quaglia-Senta (cit. 48) and Durinian (15). Durinian based his map of auricular OPAs on detailed studies of the auricular innervation. He believed that diminished impedance in OPAs is caused by constant stimulation of the convergence neuron by signals from a diseased organ (the convergence neuron is connected to both the internal organ and the related OPA). Welhover suggested that not only Head's dermatomes but also sensory organs could act as output terminals for afferent nerve signals (91).

At a certain stage, there was a lot of interest in Kirlian photography, also known as "electrography," "electrophotography," and "corona discharge photography." This is a photographic image of the forced emission of superficial electrons from the subjects under the influence of high-voltage and high-frequency electric fields. However, there is no clinical evidence that justifies the use of this phenomenon in medicine; changes in diagnostic images obtained with Kirlian techniques can be explained by changes in perspiration caused by different diseases.

As indicated, many authors have investigated the effect of organ pathology on the electrical parameters of the corresponding skin areas. Diagnostic methods based upon measurements of electrical potential, resistance, and impedance of these zones have been proposed; however, their diagnostic accuracy has not been proven and even reproducibility has not been consistent. A wide variety of measurement techniques and current parameters are used in the previously mentioned methods. It seems that the results mainly depend on perspiration, which is influenced by the patient's muscular tension, emotional condition, and skin hydration, as well as the procedure duration, environmental temperature and humidity, and the pressure of the measuring electrode (61). Furthermore, all of these methods require additional analysis to draw any conclusions. Therefore these methods did not find widespread application in contemporary medicine, and the authors' ideas did not create a unified and systematized scientific basis for the utilization of bioelectrical skin properties for organ diagnostics.

2.5.2 Investigation of Skin Electrical Potentials

Our investigations of skin electrical potentials (70, 72, 74, 75, 76) have been performed on groups of inpatients with various clinical diagnoses and groups of clinically healthy volunteers at the 4th Department of Internal Medicine, Silesian Medical University, Poland.

Using an oscilloscope (100 mohm internal resistance), a measuring electrode with a spring-mounted constant pressure Ag/AgCl dry point electrode (1 mm diameter—reflects the estimated area of an acupuncture point), and much larger various reference electrodes placed at various areas of the skin (Figure 2.7), the following tests were performed:

- Measurements of electrical potentials of four classical APs located on particular meridians and four skin spots located outside of meridians. For diseased persons, the measurements were taken on meridians related to diseased organs as well as proven healthy organs according to classical acupuncture rules. For healthy volunteers, the measurements were taken on the Large Intestine and Liver Meridians. The measurements always included the stimulating and sedative points of the respective meridians.
- Small acupuncture needles were inserted into the respective stimulating or sedative points, depending on the type of pathology, and the potentials of the remaining APs, as well as accidental skin spots, were checked again.
- Electrical potentials of needles were continuously monitored during their insertion in APs related to both diseased and healthy organs.

All measurements were done at room temperature and humidity.

Investigations revealed that skin potential values vary widely, ranging from 0–300 mV, and their polarization and values depend, among other factors, on sweat gland activity, the type of electrodes used, and the distance between the area under investigation and the reference electrode. Statistically significant differences between the values of the electrical potential of the examined APs and other skin spots were not found. There was no statistical correlation between the values of the electrical potential of APs and the condition of the respective internal organs related to the investigated APs on the basis of classical acupuncture rules.

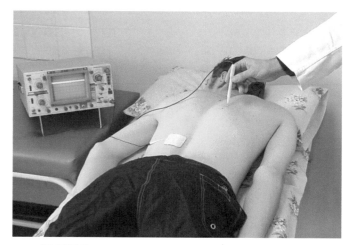

FIGURE 2.7 Investigation of skin electrical potentials.

Needle stimulation of certain APs did not change the electrical potential values of the remaining APs of the respective meridians, neither in the patients nor in the healthy volunteers.

Monitoring of the electrical potentials of needles during their insertion into APs statistically showed a sudden drop in potential of $69 \pm 23\,mV$, precisely when the so-called jump sign ("de qi") was obtained. The jump sign is a mild electric shock-like sensation (sometimes also felt as heaviness, tingling, numbness, or another feeling) that occurs when the needle is inserted correctly into an AP and is considered to be essential for therapeutic efficacy of acupuncture. The value of that potential decrease depended neither on the reference electrode material nor on the kind of APs (there was no difference seen between APs related to diseased organs and those related to healthy organs).

2.5.3 Investigation of Skin Electrical Resistance with a Standard Ohmmeter

Existing electropuncture diagnostic methods—for example, those by Voll, Nakatani, or Nogier (see Section 2.5.1)—use the measuring voltage of 1.5 V, a typical measuring voltage of standard ohmmeters. Therefore, in order to verify the foundation of these methods, our own investigations (70, 72, 75, 76) have been performed on groups of inpatients with various clinical diagnoses and groups of clinically healthy volunteers at the 4th Department of Internal Medicine, Silesian Medical University, Poland.

Using a standard ohmmeter "Meratronic-V-640," a spring-mounted constant pressure Ag/AgCl dry point measuring electrode (1 mm diameter—reflects the estimated area of an acupuncture point), and a much larger wet reference electrode (also Ag/AgCl) placed at various areas of the skin (Figure 2.8), the following tests were performed:

- Measurements of the electrical resistance of four classical APs located on particular meridians and four skin spots located outside of meridians, including the ear auricle and the lips. For diseased persons, the measurements have been taken on both meridians related to diseased organs and meridians related to proven healthy organs, according to classical acupuncture rules. For healthy volunteers, the measurements have been taken on the Large Intestine and Liver Meridians. The measurements always included the stimulating and sedative points of the respective meridians.

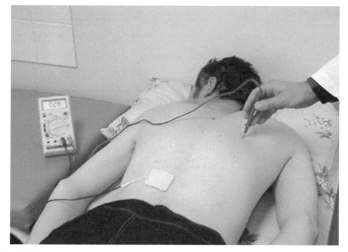

FIGURE 2.8 Investigation of skin electrical resistance using a standard ohmmeter.

- Small acupuncture needles were inserted in the respective stimulating or sedative points, depending on the type of pathology, and the resistance of the remaining APs as well as skin spots located outside meridians were checked again.
- Using another spring-mounted constant pressure Ag/AgCl dry point measuring electrode instead of the previously described reference electrode, the electrical resistance was checked for various configurations between APs located along meridians as well as for skin spots located outside meridians.

All measurements were done at room temperature and humidity.

It was found that a dense network of so-called low-resistance points (LRPs) exists on the entire body surface including the lips, where an absence of sweat glands is observed. These are skin spots of diminished electrical resistance compared to the surrounding areas. The magnitude of the electrical resistance of the LRPs equaled $393 \pm 220\,k\Omega$, and the resistance of the surrounding skin spots equaled $5.4 \pm 2.7\,M\Omega$. The difference in resistance between LRPs and the surrounding skin is greater for areas with thicker skin. The LRPs are approximately 1–2 mm apart.

Statistically significant differences between the values of electrical resistance of the examined APs and other skin spots were not found. There was no statistical correlation between the values of electrical resistance of APs (including those on the ear auricle) and the condition of the respective internal organs related to the investigated APs on the basis of classical acupuncture rules.

Insertion of needles in certain APs did not cause any changes in the electrical resistance, neither in other APs nor in skin spots located outside meridians.

The electrical resistance between APs located along the same meridian did not show any statistically significant difference in comparison to that measured between accidental skin spots.

2.5.4 Influence of Organ Pathology on the Electrical Impedance and Resistance of Organ Projection Areas of the Skin

Classical acupuncture points are believed to correspond to not just one but many internal organs/body parts. For instance, point Hegu (Large Intestine 4), located on the hand, is traditionally used not only for treatment of the local pathology but also for various diseases and disorders of the throat and head region. Point Zusanli (Stomach 36), located below the knee, is traditionally used not only for treatment of the leg pathology but also for various diseases and disorders of the stomach, intestine, and even reproductive organs.

Therefore, to estimate the influence of a particular organ's pathology on the electrical parameters of related skin areas clearly, we have not used corporal acupuncture points but have instead used organ projection areas (OPAs) of the ear auricle. It is presumed that each auricular OPA corresponds to only one internal organ/body part. Also important is that pathologies of the ear auricle are very rare, so it is very unlikely that they could influence the results of our measurements.

Patients and Methods

STUDY DESIGN/SAMPLING

An evaluation of the electrical impedance and resistance of chosen auricular OPAs related to diseased and healthy internal organs was undertaken as a double-blind study, with clinical diagnosis as the criterion standard, on 200 inpatients at the Department of Surgery, Helen Joseph Hospital, Johannesburg, South Africa (65). The group consisted of 107 men and 93 women, with a mean age of 38 years (standard deviation (SD) of 9 years).

During the postintake ward rounds, the surgical consultants preselected (in order to prevent a disproportionate number of "healthy" results versus "diseased" results) newly admitted patients with suspected pathology of one (or more) of the following organs: esophagus, stomach, gallbladder, kidneys, or urinary bladder. These organs are relatively easy to access clinically; that is, sufficient clinical data can be easily obtained in a cost-effective manner that can prove both diseased and healthy conditions. Pathologies of these organs also represent a variety of etiological and pathogenetic factors.

In each case, the evaluation of the electrical impedance and resistance of OPAs related to the previously mentioned organs (Figure 2.9) was undertaken before the final clinical diagnosis was established. The patients, sent by the independent arbiter, were always brought to the investigation room by the witness. The witness was also appointed by the independent arbiter and was either a medical doctor, a student, or a nurse. The investigator had no access to the patient's documentation whatsoever, and the witness was present during the entire measuring procedure to ensure that there was no communication between investigator and patient. The results, signed by the witness, were then handed over to the independent arbiter, who kept them in a sealed container until all investigations were completed by a separate clinical team. After the final clinical diagnosis was made, the independent arbiter identified for statistical purposes which data sets belonged to healthy organs and which belonged to diseased organs.

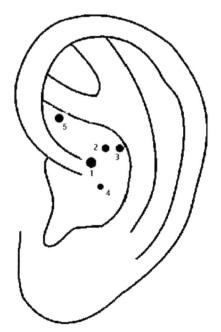

FIGURE 2.9 The locations of auricular OPAs of the stomach (1), gall bladder (2), kidneys (3), lower esophagus (4), and urinary bladder (5).

CLINICAL INVESTIGATION PROCEDURE

Clinical investigations of the chosen internal organs comprised:

1. Esophagus: history and physical examination, chest radiograph, barium swallow, and esophagoscopy with biopsy for confirmation/exclusion of esophagitis or a neoplastic process. Operative findings were included if the patient had undergone surgery.
2. Stomach: history and physical examination, barium meal, and gastroscopy with biopsy for confirmation/exclusion of mucosal inflammation or a neoplastic process. Operative findings were included if the patient had undergone surgery.
3. Gallbladder: history and physical examination, acute phase indicators, liver function tests, hepatitis markers, urine for bilirubin and urobilinogen assessment, ultrasound examination, and cholecystogram/cholangiogram (if indicated). Operative findings were included if the patient had undergone surgery.
4. Kidneys: history and physical examination, urine for microscopy, culture and susceptibility, urea and electrolytes, creatinine clearance, acute phase indicators, ultrasound examination, intravenous pyelogram, and CT scan. Cystoscopy and renal biopsy were performed if indicated. Operative findings were included if the patient had undergone surgery.
5. Urinary bladder: history and physical examination, urine for microscopy, culture and susceptibility, urea and electrolytes, creatinine clearance, acute phase indicators, ultrasound examination, and biopsy (if indicated). Operative findings were included if the patient had undergone surgery.

All clinical investigations were done in the course of normal patient care by the medical staff of the Department of Surgery, Helen Joseph Hospital. This means that a patient admitted for a stomach problem may not have undergone extensive clinical investigations of the kidneys or urinary bladder. For statistical purposes, the patient's statement that he or she did not experience any problems with these organs, supported only by physical examination, did not constitute sufficient clinical evidence to accept the condition of these organs as healthy. In terms of the double-blind trial, all clinical diagnoses were made up to the clinical team's discretion at that time. However, details of all investigations are now available in the hospital records.

ELECTRICAL IMPEDANCE EVALUATION PROCEDURE

Investigations of the electrical impedance of the previously described OPAs were made with an Ag/AgCl dry point electrode (1 mm diameter) and a larger (10 cm^2) wet reference electrode (also Ag/AgCl). In order to avoid operator bias, the point electrode was spring-mounted inside a casing such that the electrode tip applied a constant pressure (200 g/cm^2) on the skin when the casing was flush with the skin surface. The point electrode was placed on an OPA, whereas the reference electrode was placed on the patient's hand. Measurements were done using a signal generator and oscilloscope. The magnitude of skin impedance versus frequency was estimated at 5 Vpp. Measurements were taken at 10 Hz, 100 Hz, 1 kHz, and 10 kHz. The magnitude of skin impedance versus signal amplitude was estimated at 50 Hz. Measurements were taken at 1, 1.6, 2.5, 4, 6.3, and 10 Vpp. All measurements were done at room temperature and humidity.

ELECTRICAL RESISTANCE EVALUATION PROCEDURE

Investigations of the electrical resistance of the previously described OPAs were made with the same electrodes. Skin resistance versus voltage was evaluated with custom-built electronic circuitry that applied a programmable voltage (−30 V to +30 V) or current (−40 μA to +40 μA) and measured the resistance between the electrodes. All measurements were done at room temperature and humidity.

STATISTICAL PROCEDURES

An analysis of variance by means of an F test was used to check the difference between the measurement results obtained for OPAs related to healthy organs and to diseased organs (impedance and resistance results were checked separately). $p<0.05$ was accepted as the statistically significant difference. Only OPAs related to organs with proven clinical conditions (healthy/diseased) were considered for statistical purposes. Only statistical subgroups were used for final statistical calculations; casual pathologies were not included.

Results

Impedance and resistance measurement results were obtained for 1000 subjects (OPAs). However, only 538 subjects met the previously mentioned criteria. Of these subjects, 203 met the criteria for diseased organs and 335 for healthy organs.

RESULTS OF IMPEDANCE EVALUATION

Investigations of the electrical impedance of the chosen auricular OPAs indicated that a relationship exists between the skin impedance at various frequencies and the condition of the internal organ related to the investigated OPA (Figure 2.10).

The impedance of OPAs corresponding to healthy internal organs equaled $197 \pm 291\,k\Omega$ at $10\,Hz$ and decreased to $52 \pm 41\,k\Omega$ at $10\,kHz$. The impedance of OPAs related to diseased organs equaled $7.4 \pm 2.1\,M\Omega$ at $10\,Hz$ and $102 \pm 57\,k\Omega$ at $10\,kHz$. The difference in impedance for OPAs related to diseased and healthy organs was statistically significant at a level of $p<0.05$ at $1\,kHz$ and became more significant at lower frequencies (see Table 2.1). It was observed that measurements at $10\,Hz$ and to a lesser degree $100\,Hz$ produced uncomfortable sensations under the measuring electrode.

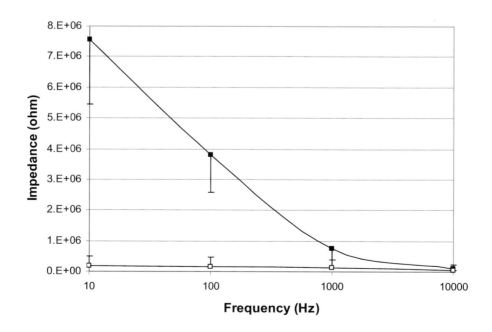

FIGURE 2.10 Skin impedance versus measurement frequency (at 5 Vpp); measurements were taken at skin zones related to diseased (solid squares) and healthy (empty squares) internal organs.

TABLE 2.1 Results of Impedance and Resistance Measurements of Organ Projection Areas (OPA) (Impedance Measured at 10 Hz, 5 Vpp; Resistance Measured at 4 Vpp with a Positively Polarized Measuring Electrode after the Breakthrough Effect)

Organ	Clinical Diagnosis	No. of Subjects	OPA Impedance Mean(SD)kohm	OPA Resistance Mean(SD)kohm
Esophagus	Healthy	50	177 (209)	258 (84)
	Esophagitis	18	6137 (1742)*	581 (389)*
	Cancer	16	9173 (1894)+	894 (292)+
Stomach	Healthy	39	264 (338)	307 (102)
	Gastritis	29	6282 (2057)*	608 (384)°
	Ulcers	17	9086 (1673)+	885 (233)*
Gallbladder	Healthy	66	242 (316)	294 (83)
	Chronic Cholecystitis	18	5659 (2121)*	439 (305)°
	Acute Cholecystitis	21	8978 (1523)+	851 (306)*
Kidneys	Healthy	108	201 (293)	275 (76)
	Pyelonephritis	14	5933 (2146)*	527 (325)°
	Hydronephrosis	26	9289 (1331)+	1005 (220)+
Urinary bladder	Healthy	72	237 (326)	282 (91)
	Cystitis	23	5821 (1979)*	511 (337)°
	Cancer	21	8813 (1440)+	842 (289)*

°Statistically significant difference in comparison with "Healthy": $p<0.05$
*Statistically significant difference in comparison with "Healthy": $p<0.01$
+Statistically significant difference in comparison with "Healthy": $p<0.001$

Investigations of skin impedance versus voltage (Figure 2.11) suggest that the disparity between the impedances of OPAs related to diseased and healthy organs is greater at higher potentials (limited by pain threshold).

There were no significant differences between the results obtained for various internal organs. The results were consistent and reproducible; they were not affected by the etiology of disease. However, there were significant statistical differences between the results obtained for generally subacute pathologies (for example, esophagitis, gastritis, chronic cholecystitis, pyelonephritis, or cystitis) and generally acute pathologies (for example, stomach ulcers, acute cholecystitis, hydronephrosis, or cancers) (see Table 2.1).

RESULTS OF RESISTANCE EVALUATION

Investigations of the electrical resistance revealed that the resistance of skin beneath the point electrode, when polarized negatively, undergoes a rapid resistance decrease of approximately two orders of magnitude (Figure 2.12) if the applied current is sufficient.

After this reversible *breakthrough effect* is obtained, the skin exhibits rectification; that is, it behaves as a diode (Figure 2.13). The degree of rectification is low for OPAs related to healthy organs. However, if the related organ is diseased, the resistance measured with a positive polarization can be five times greater than the resistance measured at the same voltage with the same but negatively polarized electrode.

The disparity between the resistance measured with a positively and with a negatively polarized electrode at OPAs related to diseased organs is greater at higher-measurement voltages; at

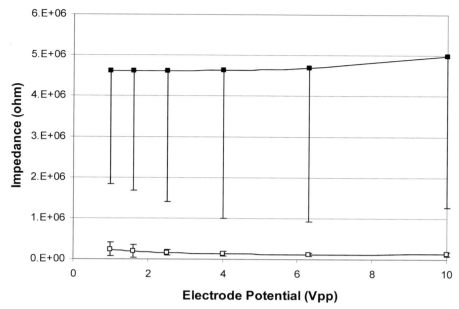

FIGURE 2.11 Skin impedance versus applied voltage (peak to peak) at 50 Hz; measurements were taken at skin zones related to diseased (solid squares) and healthy (empty squares) internal organs.

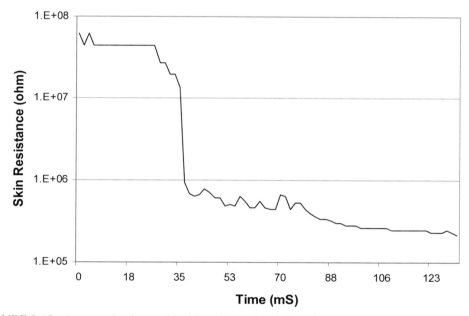

FIGURE 2.12 An example of reversible skin resistance breakthrough under constant voltage stimulation.

a measurement voltage of 0.5 V, the resistance difference was statistically significant with $p<0.01$, but at higher-measurement voltages, differences became more significant. However, it was observed that currents greater than 25 μA produced uncomfortable sensations beneath the measuring electrode.

There were no significant differences between the results obtained for various internal organs; the results were consistent and reproducible and not affected by the etiology of disease. However, there were significant statistical differences between the results obtained for generally subacute pathologies (for example, esophagitis, gastritis, chronic cholecystitis, pyelonephritis, or cystitis) and generally acute pathologies (for example, stomach ulcers, acute cholecystitis, hydronephrosis, or

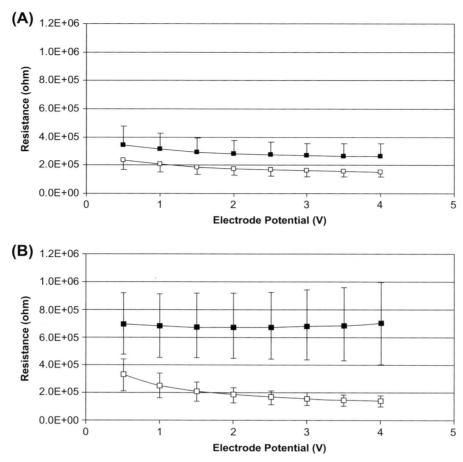

FIGURE 2.13 Skin resistance versus voltage after the breakthrough effect. A: Skin zones related to a healthy internal organ. B: Skin zones related to a diseased internal organ. Solid squares denote measurements taken with a positively polarized electrode; empty squares denote measurements taken with a negatively polarized electrode.

cancers) (see Table 2.1). Skin resistance measurements taken before the breakthrough effect did not demonstrate any statistical significance.

2.5.5 Influence of Organ Pathology on Bioelectrical Properties of Acupuncture Meridians

As detailed in Section 2.5.4, an internal organ pathology increases the electrical impedance and causes the rectification/diode effect in corresponding skin areas even if the locations of these skin areas are remote from an organ. These phenomena enabled the creation of a new method to verify the existence of acupuncture meridians as functional structures.

Our investigations (70) were performed on a group of inpatients with various clinical diagnoses and a group of clinically healthy volunteers at the 4th Department of Internal Medicine, Silesian Medical University, Poland.

Evaluations of the electrical impedance of all APs (one by one) located on particular meridians were performed with a spring-mounted constant pressure Ag/AgCl dry point electrode (1 mm diameter—reflects the estimated area of an acupuncture point) and a larger (10 cm^2) wet reference electrode (also Ag/AgCl) located on the hand or other areas of the body. Measurements were

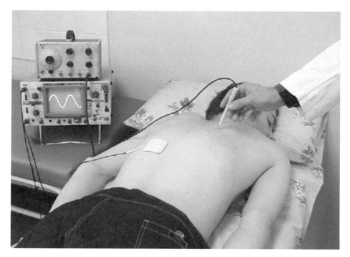

FIGURE 2.14 Evaluation of skin electrical impedance.

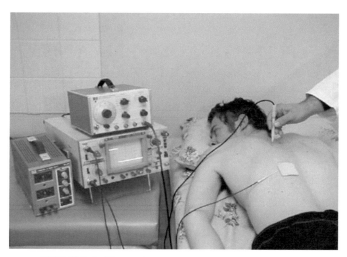

FIGURE 2.15 Evaluation of skin electrical resistance.

done using a signal generator and oscilloscope (Figure 2.14), with a measuring voltage of 5 Vpp and frequency of 10 Hz. All measurements were done at room temperature and humidity.

Evaluations of the electrical resistance of all APs (one by one) located on particular meridians were performed with an adjustable DC voltage supply, a storage oscilloscope, and the electrodes described above (Figure 2.15). An increasing voltage (maximum 20 V) was applied until the breakthrough effect was obtained, and then the resistance between the electrodes was measured for various polarizations in order to estimate the degree of rectification. All measurements were done at room temperature and humidity.

For diseased persons, investigations were performed on meridians related to diseased internal organs according to classical acupuncture rules. For healthy volunteers, investigations were performed on the Large Intestine and Liver Meridians. Once the impedance and resistance evaluation of all APs located on particular meridians was completed, small acupuncture needles were inserted in the stimulating or sedative points, depending on the type of pathology, and the evaluation of the impedance and resistance of all the remaining APs located on the investigated meridian was repeated.

The results indicated that both APs with low impedance values and a low degree of rectification and APs with increased impedance and a high degree of rectification can be found randomly on the same meridian, in both the group of patients and the healthy volunteers. This could be due to the fact that, in practice, there are no adults with perfect skeletons, and our clinically healthy volunteers could not be exceptions; perhaps corporal APs correspond not only to particular internal organs but also to the respective segments of spine and local hard tissue. Nevertheless, there was no statistical correlation between clinical diagnoses and locations of "healthy" and "diseased" points on the meridians.

The insertion of needles in certain APs influenced neither the impedance values nor the degree of rectification of the remaining APs located on the same meridian.

Our investigations of the influence of organ pathology on bioelectrical properties of acupuncture meridians did not confirm meridians as functional structures. As a by-product of these investigations, however, it was observed that all skin areas with increased impedance and higher degrees of rectification displayed tenderness (increased sensitivity to physical pressure).

2.6 INVESTIGATIONS OF THE HISTOMORPHOLOGICAL STRUCTURE OF ACUPUNCTURE POINTS AND MERIDIANS

Several authors have tried to investigate the histomorphological structures of APs and acupuncture meridians. Serebro (cit. 48, 78) found nerve structures connected with blood vessels (veins) and muscles in chosen APs. Bossy (4) obtained similar results. Bong Han (2) claimed that specific corpuscles joined by the system of tubules are present in APs. The author called it the "Kyungrak system," which was supposed to act irrespectively from other systems of the body (for example, the blood circulation system, lymphatic system, and nervous system). Kellner (29) found encysted nerve endings of Meissner and Krause's type in APs. Vandan's investigations (85), carried out with histological and histochemical methods as well as an electronic microscope, revealed the presence of subcutaneous nerves and blood and lymphatic vessels.

With regard to acupuncture meridians, some attempts were undertaken to identify them with nerves or lymphatic and blood vessels, but there are divergences between different authors (cit. 1, 20, 48, 78).

When dealing with this kind of investigation, the main problem seemed to be the lack of clear criteria of APs' localization. Our investigations therefore used specific bioelectrical characteristics of APs (see Sections 2.5.3 and 2.5.4) for precise localization. Our histomorphological investigations (10) were carried out, not later than a few hours after death, on 21 corpses at the Anatomy Department of the Silesian Medical University, Poland: 10 male and 11 female, with a mean age of 57 years (SD of 8 years).

In each case, three classical APs have been localized on the meridian related to the diseased internal organ on the basis of traditional acupuncture rules. All of these points displayed both very low electrical resistance compared to the surrounding skin area (LRPs) and a high degree rectification phenomenon (after the breakthrough effect was obtained). On the same meridian, another two skin spots were localized; these spots did not belong to classical APs, and they displayed neither low resistance compared to the neighborhood nor a high degree of rectification phenomenon.

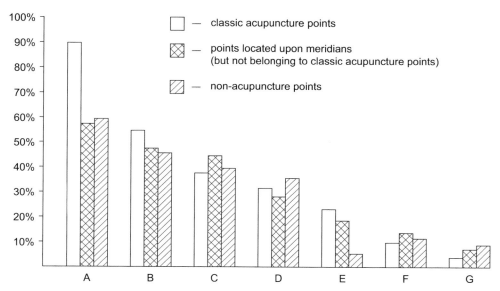

FIGURE 2.16 The presence rate of the most common histomorphological structures found in classical APs, skin spots located on meridians but not belonging to classical APs, and skin spots located outside the meridians A: free nerve endings, B: sweat glands, C: hair follicles, D: big vessels, E: encysted nerve endings, F: nerve fibers, G: sebaceous glands.

Another two skin spots have been located outside meridians; these spots also displayed neither low resistance compared to the neighborhood nor a high degree of rectification phenomenon.

In total, 147 skin samples (1–2 mm in diameter) were collected 5–10 hours after death. The samples were preserved in neutralized 10% formalin solution over a period of five days to six months. After dehydration, the preserved samples were put into paraffin and then cut in series into slides with a thickness of 15 μm. The slides were stained: 7840 with hematoxiline-eosine and 3675 with silver (Shultz modified method). The slides were examined with a luminous microscope.

The investigations revealed that only known, typical-for-the-skin histomorphological structures are present within APs and acupuncture meridians. However, APs showed a significantly higher rate of free nerve endings compared to other skin spots that did not display a diminished electrical resistance (Figure 2.16).

2.7 DISCUSSION

Reflexive physical medicine uses various forms of physical energy to stimulate and regulate the body's own powerful self-defense mechanisms and systems. This energy is usually applied to certain skin areas. Therefore, over centuries (depending on the level of scientific progress), various beliefs, hypotheses, and theories have been developed concerning both the therapeutic techniques and the skin areas to be stimulated. Some of these ideas are based on scientific research; others are less substantial.

Among various physical therapies, acupuncture is perhaps most affected by both "wishful thinking" and prejudice. The reason is that this arguably most effective and versatile system of physical medicine is also the oldest one. Having no physiological and very little anatomical knowledge, the ancient Far Eastern doctors adapted their general beliefs in the "vital energy" circulating in the human body along meridians in order to explain how acupuncture works. Surprisingly, according to them, this unmeasurable enigmatic energy could be controlled by means of such primitive tools as pieces of wood/bamboo or stone/metal needles. Chemically oriented (especially over the last seventy years), Western, academic medicine promptly rejected this "nonscientific" method because there were no obvious chemical foundations. However, some enthusiastic but often home-grown "alternative" practitioners directly recognized the hypothetical "vital energy" as electricity, laser energy, or cosmic radiation.

Significant divergences between different authors forced us to verify their sometimes controversial opinions on our own. Chapter 2 of this book summarizes the experimental part of our complex research program. Because all the sections of Chapter 2 have been published previously as separate comprehensive articles (see references), our less important research was summarized very briefly, and more detail was given for more important research.

Our thermographic investigations (see Section 2.3) confirmed neither the existence of hypothetical acupuncture meridians as energetic channels nor increased metabolism of classical APs. However, thermography can visualize TPs (ashi points), that is, skin areas that are usually localized on top of the muscle spasm which are not painful on their own but display tenderness (increased sensitivity to digital pressure) when the corresponding internal organ is diseased.

Our radioisotopic investigations of acupuncture meridians (see Section 2.4) did not provide any evidence that these hypothetical structures could be conductors of cosmic/radioactive energy.

Our measurements of the skin electrical potentials (see Section 2.5.2) showed no difference between classical APs, meridians, and other skin spots. There was also no statistical correlation between the values of these potentials and the clinical conditions of corresponding internal organs. Stimulation with acupuncture needles had no effect on the electrical potentials of other APs located along the same meridian. However, monitoring of the electrical potentials of needles during their insertion in APs revealed a statistically significant drop in these potentials when the APs were correctly punctured. The value of this potential decrease coincides with the value of inner neuron electrical potential (82, 86).

Investigation of the skin electrical resistance with a standard ohmmeter of 1.5 V measuring voltage (see Section 2.5.3) revealed that a dense network of LRPs exists on the entire body surface including the lips, where an absence of perspiration glands is observed. These are skin spots of significantly diminished electrical resistance compared to the surrounding areas. However, because the LRPs are approximately 1–2 mm apart, they cannot be regarded as classical APs. This is contrary to the claims of certain authors that only APs display lower electrical resistance. Furthermore, it means that punctoscopes, once widely advertised as precise detectors of APs, are useless; they indicate skin areas of diminished resistance practically everywhere! Our investigations also made it clear that measurements of skin electrical resistance using low measuring voltage are of no diagnostic value; all pseudodiagnostic methods of this kind (see Section 2.5.1) only serve to give acupuncture a bad name.

Our investigations into the influence of organ pathology on the characteristics of electrical impedance and resistance of APs (see Section 2.5.4) proved that OPAs, which can be identified with APs, do exist on the skin surface. Pathology of an internal organ causes significant rectification of electrical currents (diode phenomenon) in related OPAs once the resistance breakthrough effect has been induced in the skin. Pathology of an internal organ also increases the impedance of corresponding OPAs. The degree of rectification or difference in impedance is proportional to the extent of the pathological process within this organ. The disparity between the resistance measured with a negatively polarized and a positively polarized electrode at OPAs related to diseased organs, once the resistance breakthrough effect has been induced in the skin, is greater at higher measurement voltages. The disparity between the impedances of OPAs related to diseased and healthy organs is greater at lower frequencies and higher potentials. The influence of organ pathology on the electrical parameters of related OPAs does not depend on the kind of internal organ and is not affected by the etiology of pathology.

According to the previously noted phenomena, specific bioelectrical properties of OPAs/APs can only be demonstrated under the influence of an adequate external electrical source. In terms of bioelectrical features, OPAs/APs do not distinguish themselves from other areas of skin, except for the dependence of their resistance/impedance characteristics on the actual condition of respective internal organs. This specificity to particular internal organs/body parts is the only difference between APs and other skin areas.

The study proved that the skin resistance characteristics of specific locations are dependent on the state of health of corresponding internal organs. However, in contrast to previous attempts by other authors, the breakthrough effect (36, 55, 56, 60, 65, 69, 71, 72, 75, 76) was the key to obtaining skin resistance measurements that correlate with the condition of a related organ. This phenomenon has been investigated by some authors (8, 9, 21, 43, 45, 90), but never in the context of medical diagnostics. Only after it is obtained, the resistance of the diseased organs' projection areas measured by means of a positively polarized electrode significantly higher compared to the resistance for the same but negatively polarized measuring electrode. The ratio of these two measurements is not affected by all the factors that influence the actual skin resistance values; therefore, a universal point of reference is established (55, 56, 60, 65, 67–69, 72, 75, 76). The previously mentioned phenomenon of small skin areas acting as powerful electrical diodes because of damage done to a distant internal organ was never described by other authors.

The impedance of a skin area corresponding to a diseased organ is significantly higher than that of a skin area related to a healthy organ if both areas are on the same type of skin of the same individual: for example, on the ear auricle. However, skin impedance measured on different body regions—for example, the ear versus the foot—cannot be directly compared because the basic skin impedances of these regions are different (69). Once again, this kind of skin impedance dependence on the health condition of the corresponding internal organ was never described by other authors.

Our evaluation of the electrical impedance and resistance of OPAs/APs confirmed a direct functional connection between internal organs and these specific skin areas, even if these areas are distant from the organs. However, the same measurements did not confirm an existence of acupuncture meridians as functional structures (see Section 2.5.5); both APs with low impedance values as well as a low degree of rectification and APs with increased impedance as well as a high degree of rectification were randomly located along meridians. Also, the insertion of needles into the stimulating/sedative points influenced neither the impedance values nor the degree of rectification of the remaining APs located on the same meridian. The observation that all skin areas with increased impedance and a higher degree of rectification display tenderness (increased sensitivity to physical pressure) suggests that these phenomena are connected with the functioning of the nervous system.

Our investigations of histomorphological structures of APs and meridians (see Section 2.6) revealed that only known, typical-for-the-skin histomorphological structures are present within APs and acupuncture meridians. This means that no histomorphological structures have been found that could be identified with hypothetical meridians. However, APs show a significantly higher rate of free nerve endings compared to other skin spots that do not display a diminished electrical resistance. This finding directly suggests that the acupuncture therapeutic system is based on the sensory nervous system.

In general, our investigations of the physiological mechanisms, which are the basis of physical medicine, confirmed in various ways the direct functional connection between internal organs and related skin areas. They also verified various beliefs, hypotheses, and theories surrounding certain physical methods, both diagnostic and therapeutic. For example, our investigations fully confirmed the existence of APs/OPAs on the skin surface, but not the existence of hypothetical acupuncture meridians as morphological or functional structures. The meridians should be understood as artificial lines connecting APs which are roughly related to the same organs/body parts; meridians usually follow the radiation of pain originating from these organs/body parts. In practice, meridians are very useful for the localization of particular APs.

The legends surrounding certain physical methods, especially the older ones, in fact bring more harm than good to physical medicine; they create, for example, a deep division between so-called scientific, Western physical therapy and traditional, Far Eastern physical treatments. Some of the previously mentioned investigations suggest, however, that all methods of reflexive physical medicine are based on the same foundation: the nervous system.

CHAPTER

3

Neurophysiological Foundations of Reflexive Physical Medicine

OUTLINE

3.1 Introduction	33	3.4 Discussion	45
3.2 Review of Relevant Data	34	3.5 An Attempt to Visualize Organ Projection Areas	46
3.3 Convergence Modulation Theory	40		

3.1 INTRODUCTION

There is a lot of controversy surrounding the modes of action of various methods of physical medicine. Attempts have been made to provide separate explanations for all the particular types of "Western" physical therapies: for example, thermotherapy, hydrotherapy, phototherapy, ultrasoundtherapy, electrotherapy, and magnetic field therapy. However, all of these theories, which can be found in every manual of physical therapy and in the instruction manuals of various therapeutic devices, in fact produce more questions than answers.

On the other hand, traditional Far Eastern beliefs concerning so-called bio-energetic therapies have to be rejected, at least on the basis of our research described in Chapter 2.

Nevertheless, all these methods of physical medicine have much in common. For instance, they all can be called reflexive therapies, because they all stimulate the skin (sometimes mucosa) nervous receptors in order to obtain specific medical benefits not related to the skin itself. This is the most visible in the case of acupuncture, point massage, analgesic electrostimulation (TENS), laser therapy, and reflexive thermotherapy (moxa or cryotherapy). Ultrasoundtherapy, as well as shortwave and microwave diathermy, are not usually categorized as reflexive therapies. However, the reflexive mechanism may also be the leading therapeutic mechanism in these methods, as they are known to stimulate dermal nervous receptors through superficial heating (39). Of particular interest is magnetotherapy, and especially so-called pulsed magnetic field therapy with the "microTesla magnetic fields" modification, which is still therapeutically effective despite transmitting a very low energy to the body tissues (39); the sensitivity of dermal nervous receptors to changes in the electromagnetic/magnetic radiation is, however, well known. Afferent signals, sent from the skin nervous receptors to higher levels of the sensory nervous system, consist of chains of nervous active potentials. It is important to remember that the frequency of active potentials is proportional to the intensity of stimulation; these active potentials do not otherwise differ from each other, depending on the nature of the stimulus. Therefore, certain therapies can be more effective than others, but in general all methods of physical medicine, irrespective of the nature of the stimulus, produce very similar clinical effects: pain relief, relaxation of muscle spasms, improved local blood circulation, reduced inflammation and edema, and increased metabolic rate.

Many authors have tried to explain the mechanism of reflexive therapies, and accordingly, many hypotheses exist (1, 2, 7, 11, 12, 14, 15, 19, 20, 39, 44, 46–48, 52, 57, 58, 59, 78, 80, 83, 84, 93). Some of them are based on scientific research, whereas others are less substantial. However, even the most well known and outstanding theories adopted for this purpose—for example, Melzack and Wall's "gate control" theory (38, 89; Zimmermann modification, 96) and the so-called endorphinic theory (1, 7, 11, 58, 59)—show significant drawbacks in this sphere.

The "gate control" theory, as described by Melzack and Wall, can explain the "dual-phase" character of acute pain sensations; following an injury, a very brief, sharp, painful sensation occurs first, and only after a short interval is it followed by "normal," continuous pain. The authors explain this phenomenon as an interaction between the thick non-nociceptive nerve fibers ("quick conductors") and the thin nociceptive nerve fibers ("slow conductors"); both carry the information from the damaged tissue to the higher levels of the central nervous system. This specific interaction between "quick" non-nociceptive nerve fibers and "slow" nociceptive nerve fibers could be responsible for the analgesic effect of reflexive therapies. However, the gate control mechanism described by Melzack and Wall may, first, simply lead to self-blocking and, second, eliminate a superficial dermal pain perception, which is always observed during acupuncture and other more irritating reflexive therapies (for example, intensive thermotherapy or electrostimulation). In addition, Melzack and Wall pioneered the concept of gate control on the basis of their physiological observations, but never explained what would be the purpose of such a mechanism.

Endorphins and encephalins are the endogenic substances that are always released in case of any tissue damage in order to diminish pain sensations and, in this way, preserve homeostasis (82, 86). Therefore, it does not come as a surprise that acupuncture and other more irritating methods of physical medicine elevate the endorphin level in blood serum, especially after the accidental stimulation of APs/OPAs related to healthy organs. In general, however, properly applied reflexive therapies reduce the beta endorphin level in pain sufferers (12), which indicates that the analgesic effect of physical medicine is based on different mechanisms. Also, the therapeutic specificity of APs as well as their specific bioelectrical features (60, 62–65, 67, 68, 72, 75, 76) cannot be explained by the endorphinic theory.

Our research, described in Chapter 2, has confirmed a functional connection between internal organs and related skin areas directly. The dependence of OPA/AP bioelectrical features on the condition of the corresponding internal organ is particularly significant, because this phenomenon communicates important new information concerning the physiological mechanisms that are the basis of physical medicine.

3.2 REVIEW OF RELEVANT DATA

To properly discuss physiological mechanisms that could be a basis for reflexive physical medicine, all of the following facts have to be considered (71):

- When an internal organ or other particular body part is diseased, a hypersensitivity is noted at the corresponding dermatomes. TPs also appear at related skin areas, usually on top of muscle spasms. These points, which still belong to classical APs, distinguish themselves because of their increased tenderness and elevated temperature (Figures 2.3 and 2.4, Chapter 2), compared to the surrounding areas (70). Other respective APs, which are not located on larger muscles, also become more sensitive to digital pressure but do not display elevated temperature. Therefore, to distinguish them from TPs, it is practical to call these tender skin areas pressure points (PPs).
- A connection also exists between the state of health of internal organs and the electrical characteristics of related, although sometimes remote, skin areas. These skin areas are referred to as OPAs and include APs. Pathology of an internal organ causes related OPAs/APs to rectify electrical currents (55, 56, 60, 62–65, 67, 68, 72, 75, 76), once the resistance breakthrough effect (36, 65, 71, 72, 75, 76) has been induced in the skin. Pathology of an internal organ also increases the impedance of corresponding OPAs/APs (65, 72, 75, 76).
- According to histomorphological investigations (4, 10, 29, 85, cit. 20, 48, 78), APs are collections of nervous receptors—especially unsacculated free nerve endings (10). There is also other evidence that suggests the presence of nervous structures within APs: namely, the

phenomenon of a sudden decrease in the electrical potential of a needle correctly inserted into an AP (74).

- A needle inserted into an AP initially stimulates simultaneously all neighboring nervous receptors, both nociceptive—activated by various chemicals released from damaged cells—and non-nociceptive—those of palpation, extension, and temperature (Figure 3.1). As a result, for a short period of time, information concerning skin irritation is sent via both thick non-nociceptive fibers ("quick fibers") and thin nociceptive fibers ("slow fibers") to the upper levels of the central nervous system (CNS). However, after 3–5 seconds, the needle no longer causes pain sensations, and only non-nociceptive receptors still undergo stimulation, as in the case of mild AP stimulation by means of heat, cold, a weak electrical current, a low-power laser, a magnet, or massage. Nevertheless, it is well known that such procedures are therapeutically effective.
- A mild irritation of particular skin areas may arrest visceral reflexes and modify nociceptive signals from the damaged organ (see experiments with the electroacupuncture blocking of evoked potentials (16, 48, 50–52, 80); see Figure 3.2). It is remarkable that acupuncture and

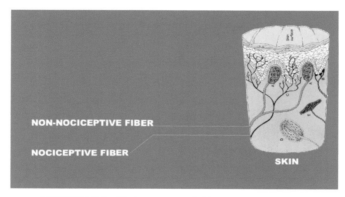

FIGURE 3.1 Various types of skin nervous receptors.

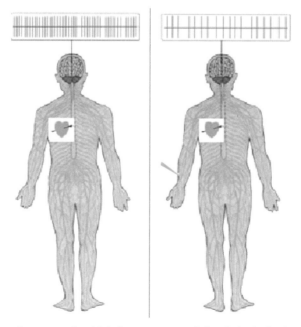

FIGURE 3.2 Damage done to an internal organ evokes high-frequency neural signals in the brain, which can be measured with microelectrodes. Even mild stimulation of specific skin areas can modify these signals.

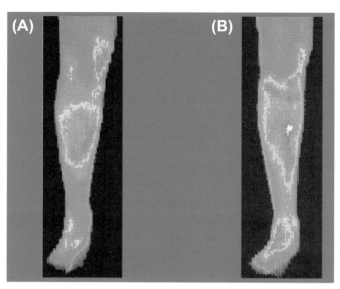

FIGURE 3.3 Thermogram of the ischaemic lower extremity before (A) and after (B) electroacupuncture treatment. The bright area reflecting a higher temperature is much larger after the treatment, indicating improved local blood circulation.

other reflexive therapies alleviate the pain caused by internal organ pathology while inducing sensations at the stimulated skin zone.
- Various methods of reflexive physical medicine stimulate dermal nervous receptors, which generate afferent signals sent to higher levels of the sensory nervous system. These signals consist of chains of nervous active potentials. The frequency of active potentials is proportional to the intensity of stimulation, but these active potentials do not otherwise differ from each other, depending on the nature of the stimulus. Therefore, certain therapies can be more effective than others; but in general, all methods of reflexive physical medicine, irrespective of the nature of the stimulus, produce very similar clinical effects: pain relief, relaxation of muscle spasms, improved local blood circulation, reduced inflammation, and edema, as well as increased metabolic rate (39). Improved local blood circulation (see Figure 3.3) may contribute to the accelerated healing of damaged tissues and improve conditions such as chronic infection/inflammation or neuropathy.
- The analgesic effect of local thermotherapy procedures (cryotherapy at −160 degrees Celsius or heating at +45 degrees Celsius) depends on both the intensity and the amplitude of the thermal stimulation (see Figure 3.4) (26, 27). The most effective electrical current in electroanalgesia is that which comprises the most "irritating" parameters: bipolar impulses with high amplitude and short duration, low frequency, and modulation of all parameters (Figure 5.7) (73). Neither perception nor pain thresholds of the skin are changed during these procedures (25, 26).
- Reflexive therapies in which a constant stimulus is used—such as constant heating or cooling, DC stimulation, static magnetic fields, continuous laser stimulation, or classic acupuncture (without needle manipulation)—are therapeutically less effective than therapies using the same stimuli in dynamic form, such as alternating heating and cooling, AC stimulation, pulsating electromagnetic fields, pulsating laser stimulation, or electroacupuncture (19, 26, 27, 39). Dynamic stimulation, especially with high amplitude and intensity (within safe limits), induces higher frequencies of afferent nervous pulses. The frequency range of these dynamic therapies is generally 1–200 Hz, which coincides with the frequency range that is best detectable by dermal receptors. Higher stimulating frequencies do not increase nervous activity (82, 86).

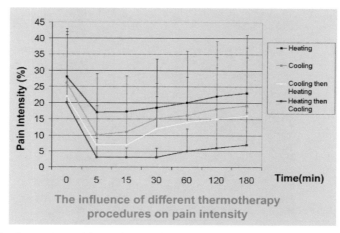

FIGURE 3.4 Comparison of the analgesic effectiveness of various thermotherapeutic procedures.

- The thalamus and hypothalamus act as a "transmission station" that induces adequate changes in the hormonal balance in response to changes in the frequency of afferent nervous impulses. The impulse frequency encodes information about the condition of the relevant internal organ (pathology intensity only; no message about the type or the etiology of the disease). Therefore, if by means of reflexive therapy some nervous information tracts are blocked, particular hormonal and humoral responses can be expected, such as changes in the serum concentration of ACTH, cortisone, adrenaline, noradrenaline, or testosterone (12, 44, 93, cit. 1, 14, 20, 39, 48, 78).
- Skin rectification/increased impedance phenomenon is connected with the sensory and not the autonomic nervous system. Measurements of the electrical rectification phenomenon of OPAs/APs after the breakthrough effect is achieved remain unchanged even though cryotherapy procedures (at −160 degrees Celsius) were carried out simultaneously (see Figure 3.5) (unpublished observations). It is well known that cold is a potent factor that influences the autonomic nervous system and thus induces cutaneous blood vessel contraction. Therefore, if the autonomic system was responsible for the nociceptive information supply from the diseased organ to the corresponding skin area, a sudden application of a cold stimulus should influence the measurement results. This did not happen.
- The CNS cannot process all available information that originates both internally and externally at the same time due to limited capacity. The need to eliminate less immediately important information created the specific structure of the sensory nervous system in which many lower level fibers converge to a single fiber at a higher level; this is the neural convergence principle (71, 82, 86) (Figure 3.6). Higher priority is given to signals resulting from external stimuli than to messages coming from internal organs, because the information coming from sensory organs, including the skin, is generally more important for the organism's self-defense and survival.
- Nociceptive signals sent from a diseased organ to the CNS can also reach related skin areas (antidromic direction), because there is a close dependence of the bioelectrical characteristics of particular skin areas, such as OPAs/APs, on the actual condition of the corresponding internal organs (55, 56, 60, 62–65, 67, 68, 72, 75, 76). Therefore, nerve tracts that carry information from damaged internal organs must be linked with nerve tracts that conduct information from the skin surface receptors. This connection, predicted by Morley (cit. 14, 82),

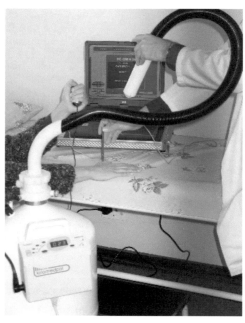

FIGURE 3.5 A cryotherapy procedure has no influence on the skin rectification/increased impedance phenomenon.

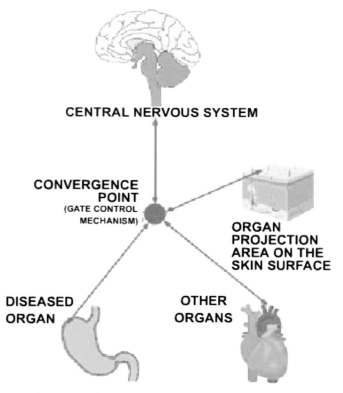

FIGURE 3.6 Neural convergence principle. Nociceptive signals sent from a diseased organ to the CNS can also reach related skin areas (antidromic direction). Signals coming from damaged/stimulated OPAs are of higher priority, so a temporary blocking of signals originating from internal organs can occur in the convergence point.

is in accordance with the convergence principle of the nervous system (82, 86). The contact point (convergence area) is most probably situated before the synapse that links the first and second neuron of the afferent tract from the AP (Figure 3.7). If the contact point was situated after this synapse, such a phenomenon would not be possible, because synapses are thought to transfer information in one direction only.

- Stimulation of sensory nerves has been shown to result in the release of neuropeptides not only at the synapse but also at the innervated epidermal area (Figure 3.7) (3, 37, 40, 82, 86). This results in vasodilation and increased capillary permeability in the vicinity of the stimulated nerve. For example, electrical stimulation of the sensory saphenous nerve of the cat leads to neurogenic edema formation in the innervated area of the paw (3). The non-nociceptive nervous receptors are sacculated, but nociceptive free nerve endings release neuromediators directly to the intercellular fluid.
- Specific bioelectrical phenomena of the skin can be seen even on unconscious patients (62) (Figure 3.8) and fresh corpses (10). The latter means that changes in the skin that are

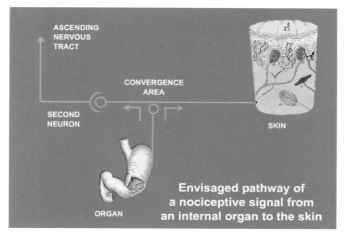

FIGURE 3.7 An envisaged pathway of a nociceptive signal from an internal organ to the skin. The release of neuropeptides takes place not only at the synapse but also at the innervated epidermal area.

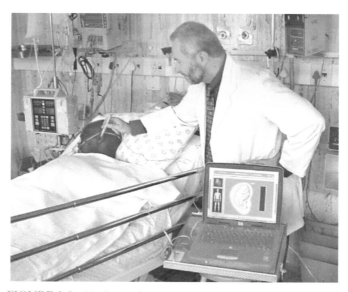

FIGURE 3.8 Bioelectrical measurements on an unconscious patient.

responsible for the appearance of the rectification phenomenon in the case of the pathology of corresponding internal organs are still preserved after death. This implies that these changes are of a structural rather than functional nature.

3.3 CONVERGENCE MODULATION THEORY

The convergence modulation theory (71) described ahead can elucidate all the facts and phenomena reviewed in Section 3.2 and in this way provides a scientific explanation of the biological action of reflexive physical medicine.

If the specific internal organ marked in Figure 3.9 as "O" underwent damage (for example, if it is a stomach with an acute ulcer), then the respective nociceptive nervous receptors would commence transmission of information regarding the tissue destruction to the corresponding second neuron "CN" in the spinal cord. This information is coded in the increment of the frequency of afferent impulses. The information is then transmitted to higher levels of the CNS. However, the local visceromotor reflex arc starts to act simultaneously (resulting in a local vasoconstriction and strong spasm of abdominal muscles—"M" in this example). In accordance with the convergence principle, the nociceptive fiber from the skin (AP) can be attached to the same second neuron CN as indicated in Figure 3.9. This type of fiber connection, namely the "addition" of the fiber arising from an organ to that of one arising from the skin, before the first synapse of the converging neuron CN allows stimulation of the entire neuron that is transferring information from the skin.

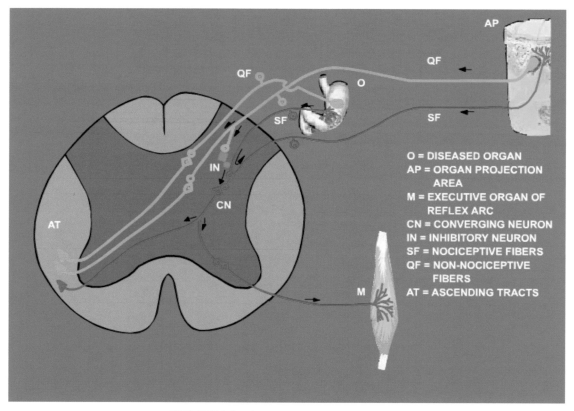

FIGURE 3.9 Convergence modulation theory.

This results in the appearance of afferent impulses at the corresponding AP, making this skin area tender. The second neuron CN, which is common to both the diseased organ and the related skin area, is unable to differentiate whether it is the organ or the skin that is affected, although it is able to provide CNS with information regarding pathology. At the same time, however, information concerning changes in the diseased organ (for example, temperature or tension) is also transmitted along thick non-nociceptive fibers "QF" arising from the affected organ O. This assists the CNS with differentiation.

During stimulation of the dermal nerve receptors (by means of needle, heat or cold, laser, magnetic fields, massage, or electrical current application), the neurophysiological mechanism that gives priority to information arising from the sensory organ—the skin (see Sections 3.2 and 3.4)—is activated. Thus, a temporary blocking of thin nociceptive fibre from the diseased organ O has to occur. This may be due to the action of the hypothetical inhibitive neuron "IN" described by Melzack and Wall (38), which is thought to be located within a gelatinous substance of the posterior horn of the spinal cord. The neuron IN would then be stimulated by the branch of thick non-nociceptive fibre QF arising from the stimulated skin area (AP). Instead of this hypothetical neuron, an inhibitory synapse may simply function here in a similar manner. AP stimulation does not exclude dermal sensitivity because the thick non-nociceptive fiber QF arising from the skin carries information such as skin temperature or pressure to higher levels of the CNS. In the actual situation, the patient experiences an improvement in pain originating from the diseased organ O while still feeling sensations caused by AP stimulation. For reflexive therapies, the perception of pressure, temperature, or touch originating from a treated organ are also not typically blocked. However, the visceromotor reflex arc is halted, as a result of the blocking of nociceptive impulses originating from the diseased organ O. This may lead to muscular relaxation with improved local blood circulation due to opening of contracted arteries and sometimes to the total disruption of the pain–spasm–pain vicious circle. In such a case, even the cause of complaints could be removed, as it is usually observed, for example, during an electroacupuncture treatment of ureterolithiasis (54, 77) when the stone can easily move down thanks to the ureteric spasm relaxation induced by the pain blockade at a local level of the spinal cord. In the case of nonpainful conditions such as asthmatic bronchospasm, the blockade of afferent impulses originating from the bronchi will also secondarily result in the spasm relaxation (79). Improved local blood circulation may contribute to the accelerated healing of damaged tissues and improve conditions such as chronic infection/inflammation or neuropathy.

Reflexive therapies control the frequency of afferent impulses reaching the thalamus and hypothalamus, thereby inducing changes in hormonal levels (12, 44, 93, cit. 1, 14, 20, 39, 48, 78). Because the leading mechanism of reflexive therapies is synaptic blocking of pain signals, a reduced level of beta endorphins will be observed in patients' blood serum after each reflexive treatment (12). In this natural way, although it is a secondary effect, the control of hormonal levels and other self-protective systems is possible.

The stimulating action of reflexive therapies on internal organs—such as the induction of contractions of an atonic bladder (19, 54, 77) or uterus (unpublished observations)—achieved by means of irritation of related skin areas is also possible by the utilization of the previously described reflexive mechanism. In this case, it is a result of the stronger stimulation of nociceptive receptors of the skin (AP) and secondary mobilization of the appropriate visceromotor reflex arc M.

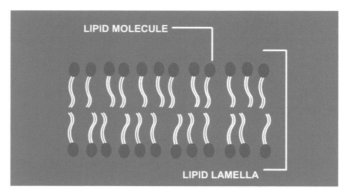

FIGURE 3.10 Molecular structure of the outermost layers of the skin.

The way in which nervous afferent impulses arising from the diseased organ cause the phenomenon of electrical rectification and increased impedance in corresponding skin areas can be clarified as follows:

- According to the theory (Figure 3.9), nervous signals originating from the damaged organ O stimulate the entire sensory nerve fiber "SF" arising from the corresponding skin area (AP). This results in the release of neuropeptides also at the innervated epidermal area (3, 37, 40, 82, 86). Increased capillary membrane permeability can therefore be expected in the vicinity of free nerve endings (3), which allows extravasation of blood plasma protein molecules (mainly albumins) (32, 90, 92). Intercellular fluid has the same composition as blood plasma, except that the concentration of proteins in blood plasma is much higher than in intercellular fluid.
- Electrical stimulation of the skin surface, such as with the measuring electrode, leads to structural changes in the stratum corneum known as electroporation (8, 9, 22). The stratum corneum consists of layers of lipid lamellae (Figure 3.10). Due to lateral thermal fluctuations of lipid molecules, pores are spontaneously formed in these lipid membranes. These pores are randomly and continuously created and destroyed. Under the influence of an electric field, however, they enter a stable state (Figure 3.11). The pores are pathways through which ions can move more freely across the membrane and therefore increase membrane conductance.
- Under sufficient electrical stimulation, the stratum corneum undergoes a rapid resistance decrease known as the reversible breakthrough effect (8, 9, 21, 36, 43, 45, 65, 71, 72, 75, 76, 90). It seems that this phenomenon is due to the previously mentioned electroporation (45, 90); the pores created in the stratum corneum allow ionic current to pass through the skin more easily. It has been shown that large molecules can act as pore blockers in electroporated bilayer membranes (8). Albumin molecules have the correct dimensions to act as pore blockers (Figure 3.12).
- The rectification phenomenon is only seen after the reversible breakthrough effect. Under these conditions, the skin resistance measured with a positively polarized electrode is always greater than the resistance measured with a negatively polarized electrode. A positively polarized electrode attracts the negatively charged albumin molecules in the extracellular fluid to the skin surface. These molecules block the electropores, reducing the conductance (Figure 3.13). A negatively polarized electrode repels the albumin molecules, thereby increasing the conductance (Figure 3.14).
- The extent of the damage done to an internal organ determines the degree of electrical rectification observed at the related skin area (55, 56, 60, 62–65, 67, 68, 72, 75, 76). The higher

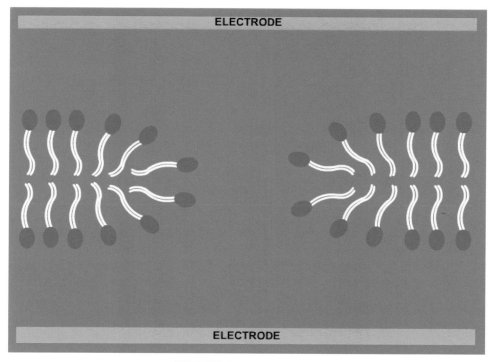

FIGURE 3.11 Electropore.

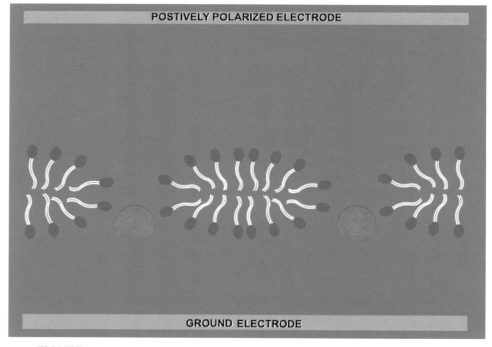

FIGURE 3.12 Albumin molecules have the correct dimensions to act as pore blockers.

the level of tissue damage, the higher the frequency of afferent nervous action potentials and the higher the amount of neuropeptides released at the related epidermal area (AP). This would result in a higher local concentration of large albumin molecules (released from capillaries), which in turn would block the electropores, thereby increasing the degree of electrical rectification observed at the skin area.

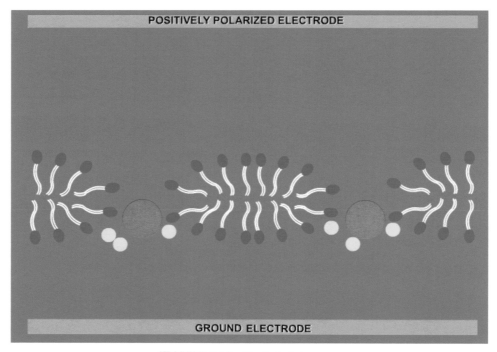

FIGURE 3.13 Blocked electropore.

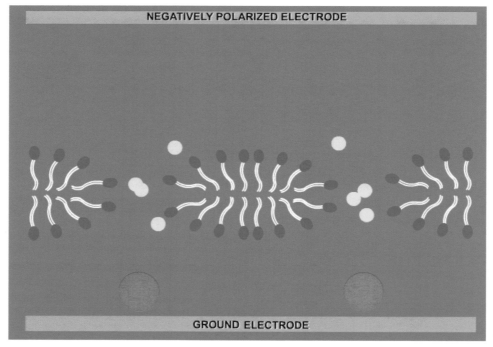

FIGURE 3.14 Open electropore.

- In contrast to DC, AC is able to pass through the skin surface via a capacitive pathway. At a high frequency, the impedance of the capacitive pathway is very low; hence the current largely flows through this pathway. At low frequency (<1 kHz), however, the voltage–current characteristic of the skin is asymmetric (43), with much lower current peaks during positive cycles than during negative cycles. This is due to lipid membrane breakdown and pore blocking by albumine molecules; this means that the higher local concentration of albumin molecules, the higher the impedance will be.

In this way, the signals sent by damaged organs to the CNS can also be detected and assessed at the skin surface.

3.4 DISCUSSION

The convergence modulation theory still requires some experimental support in certain areas. Nevertheless, it is an attempt based on contemporary scientific principles to explain all the aspects of the physiological foundations of reflexive physical medicine. It also contributes towards a better understanding of pain mechanisms and the general functioning of the nervous system: in particular, the sensory nervous system—the body's primary information network. The use of the body's potent self-defense mechanisms depends on this information. Therefore, controlling the human body's information system creates entirely new opportunities for both diagnostics and therapy.

The nervous system is the primary computing system of the human body. The sensory nervous system detects any damage done to the body, from both outside and inside, and sends the information to the decision-making centers of the CNS. The CNS cannot simultaneously process all available information that originates both internally and externally due to limited capacity. Direct nervous connections between various parts of the body and the decision-making centers of the brain would need the CNS to be physically much larger in order to sufficiently coordinate all incoming information. However, the human race would not survive with such large heads. One can draw this comparison: instead of a large stationary computer, we have a small but mobile laptop. Therefore, the need to eliminate less immediately important information created a specific converging structure of the sensory nervous system. In fact, the need to insure the priority of the most important signals originating from sensory organs might be a main reason for the existence of all these complicated anatomical structures of the spinal cord. An accidental side effect of these structures is that signals sent from internal organs to the CNS can also reach certain skin areas (OPAs/APs) and influence their electrical characteristics. In this way the OPAs/APs can act as "input/output" terminals of the body's own information network; mild stimulation of the relevant OPAs/APs can block painful sensations generated by internal organs. In addition, by breaking the relevant reflex arc, chronically contracted muscles can be relaxed and local blood circulation improved. Also, adequate changes can be induced in hormonal levels. This explains why physical medicine can be used not only for pain management but also for treatment of various disorders (such as bronchospasm, esophagospasm, spastic colon, or atonic bladder or uterus) and accelerated healing of damaged tissues (wounds, chronic inflammations, or neuropathies). *In general, reflexive physical medicine seems to function by controlling the flow of information in the nervous system and thereby reprogramming the body's powerful self-defense mechanisms and systems according to actual needs.*

In a similar way, some car manufacturers have already made practical use of the selection of information concept: The Intelligent Driver Information System (IDIS) delays passing nonessential information to the driver until it is safe to do so. For example, it will answer the phone, but it will only connect once conditions are appropriate.

The history of medicine also provides good examples to illustrate this concept. The phenomenon of soldiers still running on the battlefield with both legs already broken is well known. The battlefield provides so much vital information through the sensory organs that there is no space left to carry less important messages from internal organs to higher levels of the CNS. Other good examples include small children suffering from something such as stomach pain. Under normal circumstances, they usually continuously cry and complain; but if they suddenly see, for example, a car accident in the street, they immediately stop crying and visibly do not suffer for a certain period of time.

The opposite is also true; for example, most chronically diseased sufferers report more pain at night. This is, perhaps, due to the fact that at night, when it is quiet, not much information comes from outside, and painful signals originating from diseased body parts can more freely reach higher levels of the CNS.

Nervous receptors are known to be sensitive to electromagnetic stimulation; electromagnetic field therapy is one of the most important methods of physical medicine. The interesting phenomenon of chronic sufferers feeling better after a weather change can be explained by the sufferers' dermal receptors being stimulated by changes in the electromagnetic cosmic radiation emitted by the sun's volcanic eruptions. This "natural electromagnetic therapy" can result in temporary relief. Because weather changes follow changes in cosmic radiation, chronic sufferers usually associate changes in their conditions with changing weather.

Convergence modulation theory explains also the importance of the precise localization of the skin areas to be stimulated with local reflexive therapies. If skin areas chosen for a therapy such as acupuncture, point massage, lasertherapy, TENS, or thermotherapy (moxa or cryostimulation) are accidental ones, the afferent nervous signals from both the diseased internal organ and the stimulated skin areas will be sent via separate nervous tracts to higher levels of the CNS. As a result, the patient will experience both the pain originating from the organ and the sensations generated by the skin stimulation. In such a case, any subjective relief reported by a patient has to be regarded purely as a placebo effect; the wrong choice of the stimulated skin areas can explain why some researchers regard acupuncture and related therapies as a placebo. According to our research (see Chapter 2), all skin areas chosen for therapeutic stimulation should display tenderness (practical but subjective criterion) and high degree rectification/increased impedance phenomenon (scientific objective criterion). Tenderness of the OPAs/APs related to the diseased organ can be explained by the presence of high-frequency action potentials (originating from the organ) in the sensory nerves innervating these skin areas.

3.5 AN ATTEMPT TO VISUALIZE ORGAN PROJECTION AREAS

The convergence modulation theory (see Section 3.3) explains the phenomena of increased tenderness of OPAs/APs related to diseased internal organs and increased impedance and high degree rectification measured in these skin areas. Due to the specific structure of the sensory nervous system, nervous afferent signals sent from damaged organs to the central nervous system can also reach, in an antidromic way, the related OPAs/APs. This results in the release of neuropeptides from the local free nerve endings to the intercellular fluid of the innervated epidermal areas. Higher concentration of neuropeptides in turn causes vasodilatation and increased capillary permeability in the vicinity of free nerve endings, which allows extravasation of blood plasma protein molecules (mainly albumins). Intercellular fluid has similar composition as blood plasma, except that the concentration of proteins in blood plasma is much higher than in intercellular fluid. It is presumed that a higher local concentration of the extravasated albumins in the intercellular fluid is responsible for the previously mentioned electrical phenomena observed at the skin surface.

The hypothetical presence of neuropeptides in OPAs/APs related to diseased organs creates a unique opportunity to visualize these specific skin areas by means of the respective markers. Unfortunately, at this stage, the *in vivo* neuropeptide markers that could be detected under the human epidermis are not available. Therefore, in the meantime, another option has been explored: the search for higher albumin concentration in the respective skin areas.

In laboratory mice, albumin leak in superficial organs was traditionally followed by colorimetry or morphometry through the use of albumin-binding vital dyes such as Evans Blue (87). On our volunteers, however, it became clear that the much thicker human epidermis prevents any visibility of subcutaneous changes of this kind. A very similar problem was confronted while trying to adopt a fluorescein angiography (widely used in ophthalmology) for the same purposes.

The introduction of tagged albumin that can be detected by various imaging methods, such as magnetic resonance imaging (MRI) and positron emission tomography, opened new possibilities for quantitative, three-dimensional, dynamic analysis of permeability in any organ (87). Therefore, the contrast-enhanced MRI study of skin microvasculature, using labeled albumins, has been undertaken as an attempt to visualize OPAs/APs related to diseased organs.

The preselected comparative study was done at the Charlotte Maxeke Johannesburg Academic Hospital, on a group of volunteers with a proven either diseased or healthy condition of one of the basic organs/body parts. Volunteers were otherwise generally healthy. All of the preselected patients underwent a noninvasive bioelectrical assessment of their ear auricles (see Section 2.5.4) in order to confirm the presence of an increased impedance and high degree rectification in OPAs related to diseased organs/body parts and an absence of those electrical phenomena in auricular OPAs corresponding to proven healthy organs/body parts. Auricular OPAs have been used for this study's purposes because each auricular OPA corresponds to only one internal organ/body part, contrary to the so-called corporal OPAs (classical acupuncture points), which might be related simultaneously to quite a few internal organs/body parts. In addition, the ear auricles are separate, thin body parts that do not contain large vessels or bones. Any pathologies or direct injuries of ear auricles which could produce increased vascular permeability on their own are also very unlikely.

A contrast-enhanced MRI assessment of skin vascular permeability of the ear auricles was done by means of a Siemens Magnetom Avanto MRI machine. Special MRI sequences have been used to create a 3-D reconstruction of the ear auricles of each participant. The first assessment was done before the contrast injection. In order to dynamically visualize local capillary leaks with an increased albumin concentration, the auricular microvasculature was assessed one, two, three, four, and five minutes after a standard MultiHance MRI contrast (by Bracco (Pty) Ltd.), which transiently binds to albumins, was injected intravenously. Obtained in this way, contrast-enhanced magnetic resonance images of the diseased organ-related OPAs versus those of the healthy organ-related OPAs were then used for final comparison.

Figures 3.15 through 3.22 show exemplary images obtained five minutes after the contrast was given; it was observed that the best visibility occurred at that time. In these images, the areas of higher albumin concentration are marked as bright patches. As anticipated, there is a strong background of small vessels present, which makes it very difficult to clearly identify those spots

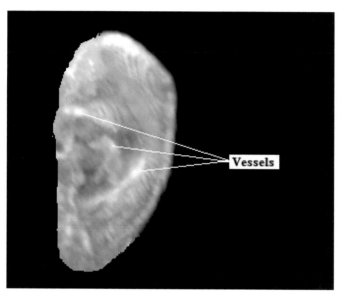

FIGURE 3.15   Contrast-enhanced MRI image of the auricular vasculature of a clinically healthy 24-year-old male (left ear auricle).

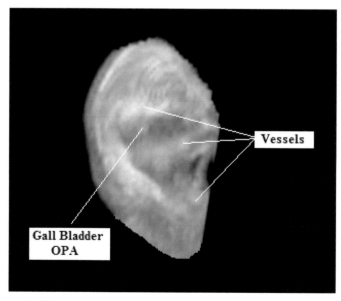

FIGURE 3.16   Contrast-enhanced MRI image of the auricular vasculature of a 29-year-old female with a gallstone (right ear auricle).

which are supposed to be OPAs; after all, the hypothetical albumin leaks are expected to be minimal compared to the albumin volumes carried by vessels. Looking at the images from different angles, however, it was possible, at least to a certain extent, to distinguish the vessels from those other respective areas of higher albumin concentration. It seems these could be the first ever images of the OPAs/APs! Nevertheless, to be certain about this we probably have to wait

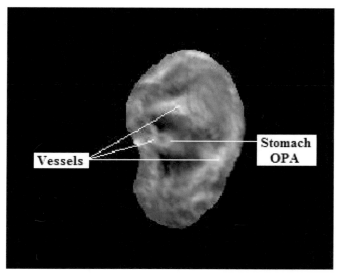

FIGURE 3.17 Contrast-enhanced MRI image of the auricular vasculature of a 33-year-old female with a stomach ulcer (left ear auricle).

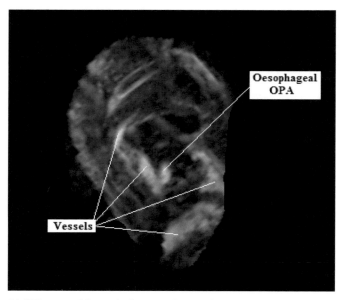

FIGURE 3.18 Contrast-enhanced MRI image of the auricular vasculature of a 33-year-old male with esophageal cancer (right ear auricle).

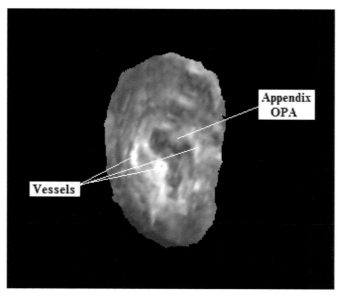

FIGURE 3.19 Contrast-enhanced MRI image of the auricular vasculature of a 30-year-old female with acute appendicitis (right ear auricle).

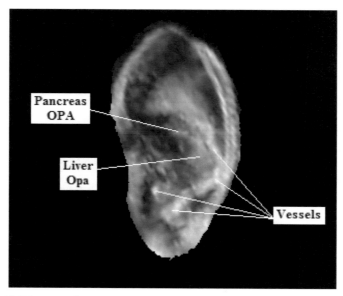

FIGURE 3.20 Contrast-enhanced MRI image of the auricular vasculature of a 51-year-old male with acute pancreatitis and alcoholic hepatopathy (left ear auricle).

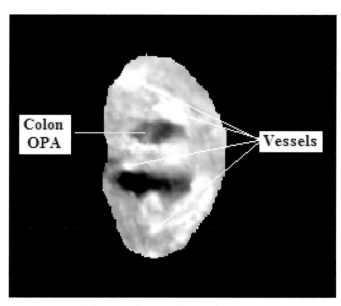

FIGURE 3.21 Contrast-enhanced MRI image of the auricular vasculature of a 47-year-old male with distal colon cancer (left ear auricle).

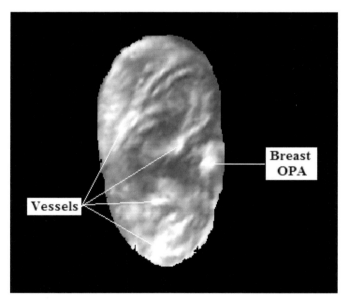

FIGURE 3.22 Contrast-enhanced MRI image of the auricular vasculature of a 39-year-old female with right breast cancer (right ear auricle).

until the suitable neuromediator markers are available. Such markers would eliminate the current problem of background interference, and in this way we should finally produce very clear, proven images of the OPAs/APs.

CHAPTER 4

Organ Electrodermal Diagnostics

OUTLINE

4.1 Implementation of Optimal Measuring Parameters for an OED Device   55

4.2 Localization of Particular Organ Projection Areas   56
   4.2.1 Localization of Auricular Projection Areas of the Stomach and Duodenum and their Use in the Monitoring of Ulcer Disease   57
   4.2.2 Localization of the Auricular Projection Area of the Liver and its Use in the Monitoring of Viral Hepatitis   60
   4.2.3 The Influence of General Anesthesia and Surgical Intervention on the Electrical Parameters of the Auricular Organ Projection Areas   61

4.3 Clinical Assessment Of OED Accuracy   63
   Patients and Methods   63
      Study Design / Sampling   63
      Clinical Investigation Procedure   64
      OED Examination Procedure   65
      Statistical Procedures   65
   Results   66

4.4 Discussion   70
   4.4.1 Innovative Aspects of OED   70
   4.4.2 Prospective Applications for OED in Contemporary Medicine   71
   4.4.3 Analysis of the Risk Associated with OED   72

Findings concerning the relationships between the state of health of internal organs and the skin's bioelectrical characteristics (see Section 2.5.4) created the groundwork for a new noninvasive diagnostic method: organ electrodermal diagnostics (OED) (55, 56, 60, 62–64, 67, 68). The location of the skin zone at which a high degree of rectification (after the skin breakthrough effect has been obtained) and increased impedance are observed indicates which particular organ is diseased. The degree of rectification and difference in impedance indicate the extent of the pathological process within this organ.

4.1 IMPLEMENTATION OF OPTIMAL MEASURING PARAMETERS FOR AN OED DEVICE

The statistical characterization of the electrical impedance and resistance of the OPAs (see Section 2.5.4) allowed the selection of optimal parameters for OED that maximize the difference in impedance or resistance measured at OPAs related to diseased and healthy organs:

- For AC measurements, low frequency and high amplitude are most suitable. Therefore, 250 Hz was selected as the measurement frequency, because lower frequencies produced uncomfortable sensations under the measuring electrode. Measuring current amplitude was chosen to be 25 μA (peak) because this was observed to be below the perception threshold.
- For DC measurements, the highest amplitude of the measurement stimulus that does not cause uncomfortable sensations is the most suitable. After the breakthrough effect, the skin resistance is very low; therefore, 25 μA was chosen as optimal.

According to these parameters, the OED device "Diagnotronics" (Figure 4.1) (68) was built in terms of the South African government's program SPII No I 200011051. This device can perform diagnostic measurements using DC or AC modalities.

When using the DC mode, a dry brass point electrode (1 mm diameter) is placed on a particular skin spot and a brass reference electrode (an area about $2\,cm^2$ is covered with a conductive gel) is placed on the patient's hand. The point electrode is mounted on a spring to ensure that the pressure applied is constant and repeatable. The point electrode is polarized negatively and the potential increased until the breakthrough effect is observed (usually between 7 V and 15 V). The current is then adjusted to 25 μA and the skin resistance measured. The polarity of the point electrode is inverted (set to the same voltage at which the skin resistance measurement was taken, but positively polarized) and a second resistance measurement is taken. These two measurement values, taken from the same skin point, are used to calculate the final result using the formula:

$$\text{Rectification ratio} = \left[1 - \frac{\text{(First Measurement)}}{\text{(Second Measurement)}}\right] * 100\,\%$$

This measurement regimen (69) is performed automatically by the Diagnotronics device; the final rectification ratio reflects the extent of the pathological process activity in the organ related to the examined OPA (normal range: 0–60%).

In the case of thick, dry skin (for example, in the plantar region) for which obtaining the breakthrough effect is more difficult, an AC-based modality is more suitable thanks to the better

FIGURE 4.1 OED examination by means of a Diagnotronics device. The locations of the skin areas corresponding to the examined organs and the diagnostic results after the examination is completed are displayed on the screen.

penetration of AC through condenser-like tissues. However, the use of impedance measurements for organ-diagnostic purposes requires separate calibration for different kinds of skin. Therefore, when using the AC mode, a $2\,cm^2$ calibration electrode covered with a conductive gel is placed in the vicinity of the OPA to be investigated and the reference electrode (matching that used for the DC mode) is placed on the patient's hand. The device determines the impedance value at 250 Hz and uses it as a point of reference to be compared with the impedance value measured with the point electrode (matching that used for the DC mode) (69). This is to avoid problems caused by individual basic skin impedance, which is different for each person and even for each part of the body. The final result is displayed in a similar way as for DC measurements: as a percentage value indicating the extent of pathological process activity in the organ related to the examined OPA (normal range: 0–60%).

The Diagnotronics device displays the locations of skin areas corresponding to the examined organs and final results on a screen as "Healthy" (0–40%), "Within normal limits" (41–60%), "Subacute" (61–80%), or "Acute" (81–99%). A printing function is also included with the device to obtain printouts of the final diagnostic results. On the basis of both clinical and technical documentation, the Diagnotronics OED system is CE certified (C52113) as the first medical diagnostic device of this kind.

4.2 LOCALIZATION OF PARTICULAR ORGAN PROJECTION AREAS

Many OPA/AP maps are currently used in physical medicine. However, significant differences exist between maps prepared by different authors (15, 20, 34, 35, 42, 48, 77, 78).

Pathology of a particular organ causes related skin areas to rectify electrical currents once the resistance breakthrough effect has been induced in the skin. Also, the impedance of skin areas corresponding to diseased organs is increased. By measuring the degree of rectification or the difference in impedance of various skin areas in diagnosed patients, the precise location of a particular OPA can be determined. This method has been used to establish the evidence-based map of auricular OPAs (Figure 4.2). The following sections illustrate the way in which this map was created.

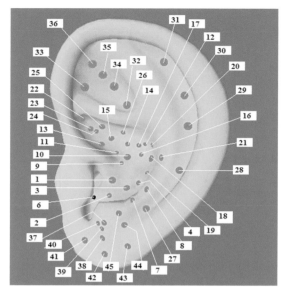
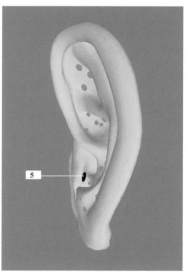

FIGURE 4.2 The evidence-based map of the auricular OPAs. The pathology of the left-sided internal organs/body parts usually causes slightly higher degrees of the respective electrical phenomena on the left ear auricle compared to the right one, and the pathology of right-sided organs/body parts usually causes slightly higher degrees of the respective electrical phenomena on the right ear auricle compared to the left one **Chest and neck**: (1) heart,* (2) bronchi,* (3) lungs* (lower lobes), (4) esophagus,* (5) breasts* (internal aspect of Tragus), (6) thyroid gland,* (7) cervical spine,* (8) thoracic spine.* **Abdomen**: (9) cardia,* (10) stomach* (pylorus), (11) duodenum* (bulb), (12) distal duodenum,* (13) small intestine,* (14) proximal colon, (15) colon* (appendix), (16) pancreas,* (17) gallbladder,* (18) liver,* (19) spleen, (20) kidneys,* (21) lumbo-sacral spine.* **Pelvis**: (22) prostate,* (23) uterus,* (24) ovaries and adnexa uteri,* (25) urinary bladder,* (26) ureters.* **Upper limbs**: (27) shoulders, (28) elbows,* (29) wrists,* (30) metacarpi,* (31) fingers.* **Lower limbs**: (32) hips,* (33) knees,* (34) ankles,* (35) metatarsi,* (36) toes.* **Brain**: (37) pons, (38) thalamus, (39) hypothalamus, (40) hypophysis, (41) cortex frontal lobe, (42) midcortex, (43) cortex posterior lobe, (44) cerebellum, (45) medulla oblongata * Location clinically proven at the time of publication *(Copyright J.Z. Szopinski, 2013.)*

4.2.1 Localization of Auricular Projection Areas of the Stomach and Duodenum and their Use in the Monitoring of Ulcer Disease

A comparative study of skin resistance characteristics over the auricular concha region in patients with stomach or duodenal ulcers and clinically healthy controls (63) was performed at the 4th Department of Internal Medicine, Silesian Medical Academy, Poland. The group of inpatients with gastric ulcers consisted of 18 people aged 36 years (standard deviation of seven years) and included 10 males and eight females. The group of inpatients with duodenal ulcers consisted of 17 people aged 33 years (standard deviation of six years) and included nine males and eight females. The diagnosis was based on history, physical examination, and the results of a gastroduodenoscopy. In all of the patients, apart from the ulcers found, there was evidence of chronic inflammation of the gastric mucosa (confirmed with a biopsy). The control group consisted of 17 clinically healthy volunteers aged 34 years (standard deviation of six years) who showed no pathological changes on endoscopic examination.

On all of the participants, evaluations of the electrical resistance characteristics of the auricular concha region were conducted with a Diagnotronics OED device. Measurements were taken on a grid of points approximately 2 mm apart, covering the whole surface of the cymba conchae. The measurements were performed on patients after the first gastroduodenoscopy (ulcers and gastritis/duodenitis present) and repeated after the final gastroduodenoscopy (healed ulcers and reduced inflammation). In the control group, the resistance measurements were done only once, after gastroduodenoscopy.

Evaluation of the electrical resistance of the auricular concha, conducted on 18 patients with gastric ulcers and gastritis before treatment, proved in each case the existence of an area of approximately 2–3 mm in diameter in which a high degree of rectification was observed. This area is located at the bottom of the cymba conchae, next to the end of crus helices (Figure 4.3).

The rectification ratio in this area equaled 87±8%, whereas the rectification ratios in the remaining examined areas equaled 37±9% (Figure 4.4). Statistically, there was no other spot in the examined region that displayed a rectification ratio higher than 60%. After successful treatment, the rectification ratio in the area described fell to 58±12%. The rectification ratios in other areas did not change significantly ($p > 0.05$).

Evaluation of the electrical resistance of the auricular concha, conducted on 17 patients with duodenal ulcers and duodenitis before treatment, proved in each case the existence of an area of approximately 2 mm in diameter in which a high degree of rectification was observed. This area is located at the upper level of cymba conchae, next to the anthelix (Figure 4.3). The rectification

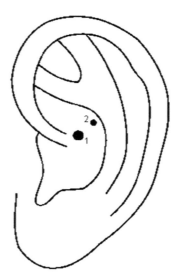

FIGURE 4.3 The location of the auricular projection areas of the stomach (1) and duodenum (distal) (2).

ratio in this area equaled 84±9%, whereas the rectification ratios in remaining examined areas equaled 41±11% (Figure 4.4). Statistically, there was no other spot in the examined region that displayed rectification ratios higher than 60%. After successful treatment, the rectification ratio in the area described fell to 52±14%. The rectification ratios in the other areas did not change significantly (p>0.05).

In the control group, no areas with a high degree of rectification were observed, and the rectification ratios equaled 42±8%. The bioelectrical measurements of the skin did not result in any observed side effects.

The study determined the precise location of the stomach projection area. However, the duodenum is a relatively long organ that crosses the abdominal cavity. For this reason, one should talk about an extended duodenal zone rather than a point; for practical purposes, at least two duodenal spots—relating to either end of the duodenum—should be considered. The first one, described in this study, corresponds to the distal duodenum. The second one, corresponding to the duodenal bulb, is located at the bottom of the cymba conchae, next to the stomach projection area (in the direction towards the eyes). Its location was proven with a double-blind clinical trial (60). Current study is concentrated on the distal duodenum OPA in order to avoid a risk of the measurement results being misinterpreted due to a closed neighborhood of the duodenal bulb and stomach OPAs (all of the patients, including those with duodenal ulcers, also suffered from chronic gastritis).

It was concluded that the existence of peptic ulcers causes the resistance characteristics of the stomach and duodenum OPAs to display a higher degree of rectification once the skin resistance breakthrough effect is induced. The results confirmed that OED measurements of these skin areas allow the noninvasive monitoring of ulcer disease.

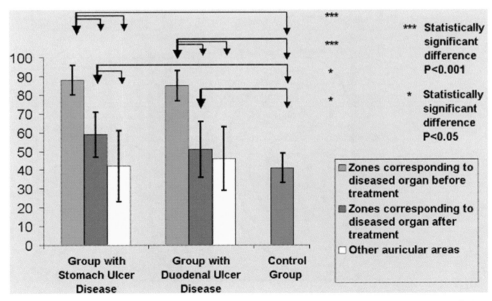

FIGURE 4.4 Comparison of the rectification ratios of various auricular regions in successive stages of ulcer disease.

4.2.2 Localization of the Auricular Projection Area of the Liver and its Use in the Monitoring of Viral Hepatitis

A comparative study of skin resistance characteristics over the auricular concha region in patients with viral hepatitis and clinically healthy controls (64) was performed at the Department of Infectious Diseases, Provincial Teaching Hospital, Tychy (Poland). The group of inpatients with hepatitis B consisted of 19 people aged 37 years (standard deviation of nine years) and included nine males and 10 females. The diagnosis was based on history, physical examination, ultrasound examination, and laboratory tests (hepatitis markers, serum transaminases, bilirubin, alkaline phosphatase, etc.). None of the patients presented with clinical symptoms of other diseases. The patients had been admitted to the hospital at least two weeks prior to the so-called "peak period": the appearance of the highest levels of serum bilirubin and transaminases. The control group consisted of 15 clinically healthy volunteers aged 35 years (standard deviation of seven years) and included seven males and eight females.

On all of the participants, evaluations of the electrical resistance characteristics of the auricular concha region were conducted with a Diagnotronics OED device. Measurements were taken on a grid of points, approximately 2 mm apart, covering the whole surface of the cymba conchae. The patients were examined immediately after admission to the hospital and then reexamined every week thereafter. In the control group, the resistance measurements were taken only once.

Evaluation of the electrical resistance of the auricular concha conducted on patients with viral hepatitis proved in each case the existence of an area of approximately 2–3 mm in diameter in which a high degree of rectification was observed. This area is located within the region of the cymba conchae, next to the anthelix and the cavity of the concha (Figure 4.5). The rectification ratio in this area, measured at the time of the peak period, equaled 82±12%, whereas the rectification ratios in the remaining examined areas equaled 48±17%. Statistically, there was no other spot in the examined region that displayed a rectification ratio higher than 60%.

In Figure 4.6, the levels of serum bilirubin and ALT are presented together with the values of rectification ratios obtained at the same time. The highest levels of bilirubin and ALT as well as

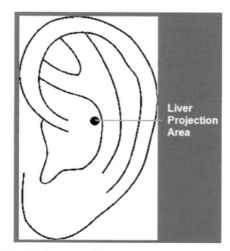

FIGURE 4.5 The location of the auricular projection area of the liver.

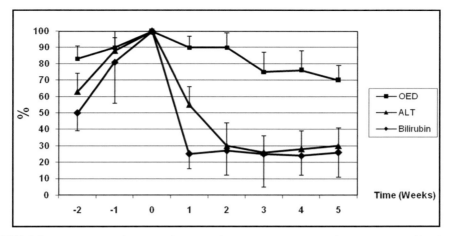

FIGURE 4.6 Comparison of OED results with Bilirubin concentration and ALT activity in blood serum.

the highest value of rectification ratio found in the liver projection area were assumed to be 100%. The peak period is marked as "0" on the coordinate axis.

In the control group, no areas with a high degree of rectification were observed, and the rectification ratios equaled 43±9%. The bioelectrical measurements of the skin did not result in any observed side effects.

The study determined the precise location of the liver projection area. The existence of viral hepatitis causes the resistance characteristics of the liver's OPA to display a higher degree of rectification once the skin resistance breakthrough effect has been induced. The results confirmed that OED measurements of this skin area allow for the noninvasive monitoring of viral hepatitis.

4.2.3 The Influence of General Anesthesia and Surgical Intervention on the Electrical Parameters of the Auricular Organ Projection Areas

To confirm the electrical rectification phenomenon experimentally and to determine the influence of state of consciousness, a comparative study of the influence of general anesthesia and internal organ damage on the electrical rectification phenomenon at related OPAs (62) was performed on a group of 30 inpatients at the Department of Surgery, Provincial Teaching Hospital, Tychy (Poland) and 15 inpatients at the Department of Gynecology and Obstetrics, Coronation Academic Hospital, Johannesburg (South Africa). The surgical group consisted of 14 men and 16 women of a mean age of 48 years (standard deviation of seven years). Sixteen patients were admitted for cholecystectomy due to gall stones, eight patients were admitted for appendectomies, and six were admitted for partial gastrectomies due to early stomach cancer. The gynecological group mean age was 25 years (standard deviation of four years). The patients were admitted for dilation and curettage (D&C) after miscarriage.

The initial value of skin resistance rectification (after the breakthrough effect) was estimated for each patient at the auricular OPA corresponding to the diseased organ (see Figure 4.7). The same procedure was performed on two control OPAs corresponding to healthy organs (heart and lungs).

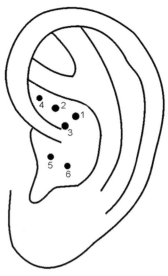

FIGURE 4.7   The locations of the auricular organ projection areas related to the gallbladder (1), appendix (2), stomach (3), uterus (4), heart (5), and lungs (6).

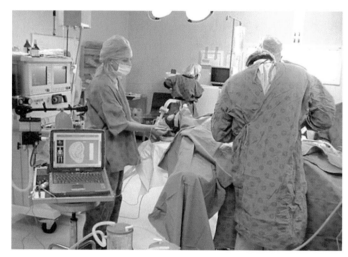

FIGURE 4.8   Monitoring of the electrical changes at the organ projection areas during a surgical operation using the OED Diagnotronics device.

A Diagnotronics OED device was used to take the measurements (Figure 4.8). Initial OED measurements taken before premedication were followed by measurements after premedication, under general anesthesia (halothane), after the skin incision, and during surgical manipulation of the diseased organ.

OED readings taken from OPAs related to diseased organs before premedication were approximately three times higher than readings from control points (see Table 4.1). Premedication, general anesthesia, and skin incision did not influence the results. Direct surgical manipulation of the diseased organs resulted in a rapid and statistically significant increase in the degree of rectification observed in the related skin areas, however.

The study directly confirmed that OPAs do exist on the skin surface. Damage done to an internal organ causes an immediate increase in the degree of electrical rectification exhibited by the related skin areas. State of consciousness does not affect this phenomenon.

TABLE 4.1 Organ Electrodermal Diagnostic (OED) Results During Successive Stages of General Anesthesia and Surgical Intervention

		Rectification Ratio (%)				
Examined OPA	Number of OPA	Before Premedication A	After Premedication B	Under General Anesthesia C	After Skin Incision D	After Surgery E
Gallbladder	16	71 ± 5	72 ± 5	72 ± 5	72 ± 5	91 ± 7#
Appendix	8	86 ± 5	86 ± 4	87 ± 5	87 ± 4	94 ± 4#
Stomach	6	69 ± 4	68 ± 3	70 ± 3	71 ± 4	95 ± 5#
Uterus	15	80 ± 6	81 ± 5	81 ± 6	–	91 ± 5#
Total	45	76 ± 8 *	78 ± 8 *	77 ± 8 *	76 ± 8 *	92 ± 6 *
Control	90	23 ± 13	21 ± 13	23 ± 14	23 ± 14	23 ± 14

#Statistically significant difference (p<0.001) in comparison with A, B, C, D.
*Statistically significant difference (p<0.001) in comparison with control.

4.3 CLINICAL ASSESSMENT OF OED ACCURACY

Section 4.1 described the technical preparation of the suitable OED device. Section 4.2 described the way in which the evidence-based map of auricular OPAs was created. With both the machine and the map being ready, it was possible to estimate the new method's diagnostic accuracy.

Patients and Methods

Study Design / Sampling

Two separate double-blind comparative studies of OED results and clinical diagnoses, as a criterion standard, were performed at the 4th Department of Internal Medicine, Silesian Medical University, Katowice (Poland) and at the Department of Surgery, Helen Joseph Academic Hospital, Johannesburg (South Africa). The first study (60) was done on the group of 230 inpatients, including 102 men and 128 women of a mean age of 43 years (standard deviation of 17 years). The second study (68) was done on a group of 200 inpatients, including 107 men and 93 women of a mean age of 38 years (standard deviation of nine years).

In both studies, during the postintake ward rounds the consultants in charge had preselected newly admitted patients (in order to prevent a disproportionate number of "healthy" results versus "diseased" results) with a suspected pathology of one (or more) of the following organs: lungs and bronchi (first study), esophagus (second study), stomach (both studies), duodenum (first study), gall bladder (both studies), pancreas (second study), colon (second study), kidneys (both studies), urinary bladder (both studies), and prostate (second study). These organs are relatively easy to access clinically; that is, sufficient clinical data can be easily obtained in a cost-effective manner that it can prove both diseased and healthy conditions. Pathologies of these organs also represent a variety of etiological and pathogenetic factors: infections, inflammation, neoplasms, and immunological and metabolic disorders.

In each case, the OED examination of the previously mentioned organs was undertaken before the final clinical diagnosis was established. The patient was always brought to the OED examination room by the witness. The witness was appointed by the independent arbiter and was either a medical doctor, a student, or a nurse. The OED investigator had no access to the patient's documentation whatsoever, and the witness was present during the entire OED examination procedure to ensure that there was no communication between the investigator and the patient. The documented OED results, signed by the witness, were handed over to the independent arbiter, who kept them in a sealed container until the final clinical diagnosis was made by a separate clinical team.

Clinical Investigation Procedure

Clinical investigations of the chosen internal organs comprised:

1. Lungs and bronchi: history and physical examination, chest radiograph, spirometry, blood gases, acute phase indicators, sputum examination. CT scan and bronchoscopy, if indicated.
2. Esophagus: history and physical examination, chest radiograph, barium swallow, and esophagoscopy with biopsy for confirmation/exclusion of oesophagitis or a neoplastic process. Operative findings were included if the patient had undergone surgery.
3. Stomach: history and physical examination, barium meal, and gastroscopy with biopsy for confirmation/exclusion of mucosal inflammation or a neoplastic process. Operative findings were included if the patient had undergone surgery.
4. Duodenum: history and physical examination, barium meal, and duodenoscopy with biopsy for confirmation/exclusion of mucosal inflammation or a neoplastic process. Operative findings were included if the patient had undergone surgery.
5. Gallbladder: history and physical examination, acute phase indicators, liver function tests, hepatitis markers, urine for bilirubin and urobilinogen assessment, ultrasound examination, cholecystogram/cholangiogram (if indicated), and hepatic immuno-diacetic acid (HIDA) cholescintigraphy (if indicated). Operative findings were included if the patient had undergone surgery.
6. Pancreas: history and physical examination, serum and urine amylase, blood glucose, fecal fats, ultrasound examination, abdominal radiograph, computed tomography (CT) scan, and endoscopic retrograde cholangiopancreatography (ERCP). Operative findings were included if the patient had undergone surgery.
7. Colon: history and physical examination, full blood count, barium enema, sigmoidoscopy and/or colonoscopy, liver function test, liver ultrasound examination, and CT scan (if indicated). Operative findings were included if the patient had undergone surgery.
8. Kidneys: history and physical examination, urine for microscopy, culture and susceptibility, urea and electrolytes, creatinine clearance, acute phase indicators, ultrasound examination, intravenous pyelogram. CT scan, cystoscopy and renal biopsy were performed if indicated. Operative findings were included if the patient had undergone surgery.
9. Urinary bladder: history and physical examination, urine for microscopy, culture and susceptibility, urea and electrolytes, creatinine clearance, acute phase indicators, and ultrasound examination. CT scan, cystoscopy and biopsy were performed, if indicated. Operative findings were included if the patient had undergone surgery.
10. Prostate: history and physical examination, urine for microscopy, culture and susceptibility, urea and electrolytes, creatinine clearance, acute phase indicators, ultrasound examination, and biopsy (if indicated). Operative findings were included if the patient had undergone surgery.

All clinical investigations were done in the course of normal patient care. This means that, for example, a patient admitted for a stomach problem may not have undergone extensive clinical

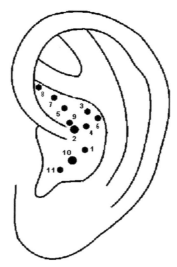

FIGURE 4.9 The locations of the auricular organ projection areas of the esophagus (1), stomach (2), gallbladder (3), pancreas (4), colon (5), kidneys (6), urinary bladder (7), prostate (8), duodenum (bulb) (9), lungs (lower lobe) (10), and bronchi (11).

investigations of the prostate or colon. For statistical purposes, the patient's statement that he or she did not experience any problems with these organs, supported only by physical examination, did not constitute sufficient clinical evidence to accept the conditions of these organs as healthy.

OED Examination Procedure

OED examinations were performed by means of the OED prototype device (55, 56) (first study) and the Diagnotronics device (second study). The examination entailed placement of the reference electrode on any area of the patient's skin, for example, on a hand, and the placement of the measurement electrode on the skin area corresponding to the appropriate organ (Figure 4.9).

The OED devices specified the actual condition of the organ related to the investigated skin area as "Healthy," "Within normal limits," "Subacute," or "Acute." A special display graded according to the percentage of the disease intensity made it possible to specify accurately the activity of the organ pathology. The locations of skin areas corresponding to the examined organs and final results were displayed on a screen.

Statistical Procedures

The comparisons of clinical diagnoses and OED results were undertaken by independent arbiters who were not involved in the diagnostic procedures.

OED detection rate was defined as the proportion of correctly classified subjects among all subjects (18, 81). Thus, the detection rate was estimated with the following formula:

$$DR = \frac{a+d}{a+b+c+d} \times 100\%$$

DR = detection rate; a = true positive; b = false positive; c = false negative; d = true negative.

OED sensitivity was defined as the proportion of correctly classified positives, that is, "true positives" among all diseased persons (18, 81). Thus, the sensitivity rate was estimated with the following formula:

$$S = \frac{a}{(a+c)} \times 100\%$$

S = sensitivity rate; a = true positive; c = false negative.

OED specificity was defined as the proportion of "true negatives" among the total of persons free of disease (18, 81). Thus, the specificity rate was estimated with the following formula:

$$SP = \frac{d}{(b+d)} \times 100\%$$

SP = specificity rate; d = true negative; b = false positive.

The positive predictive value (18, 81) of the OED result was defined as the probability of having the disease among the group of persons classified as positive by OED:

$$PV(pos) = \frac{a}{(a+b)} \times 100\%$$

PV = predictive value rate; a = true positive' b = false positive.

The negative predictive value (18, 81) of the OED result was defined as the probability of not having the disease among the group of persons classified as negative by OED:

$$PV(neg) = \frac{d}{(c+d)} \times 100\%$$

PV = predictive value rate; d = true negative; c = false negative.

The χ^2 test was used to calculate statistical significance. $P<0.05$ was accepted as the statistically significant difference. Only organs with proven clinical conditions (healthy or diseased) were considered for final statistical comparison.

Results

The OED examinations were performed on 1,610 diagnostic subjects (internal organs) in the first study and 1,600 in the second study. Sufficient clinical data to be accepted by the independent arbiters as clinically proven were available for only 516 of these organs in the first study and 714 in the second study, however.

In the first study, 463 true OED results in total were obtained from the 516 subjects considered (see Table 4.2). In the second study, 630 true OED results in total were obtained from the 714 subjects considered (see Table 4.3).

There were no significant differences in the results obtained from various internal organs (60, 68); that is, the OED results were not affected by the particular type of an internal organ. Comparison

TABLE 4.2 Comparison of Clinical Diagnoses and OED Results Obtained by Means of the OED Prototype Device ($0°$—"Healthy," $I°$—"Within Normal Limits," $II°$—"Subacute," $III°$—"Acute") at the 4th Department of Internal Medicine, Silesian Medical University, Poland

Organ	Clinical Diagnosis	No. of Subjects	True OED Results									False OED Results							
			Negative				Positive				Negative				Positive		
			0	$I°$	Together	$II°$	$III°$	Together	0	$I°$	Together	$II°$	$III°$	Together
Lungs and Bronchi	Healthy	123	36	67	103										16	4	20
	Pneumonia	28				11	16	27		1	1						
	Asthma	24				12	10	22		2	2						
Stomach	Healthy	21	8	9	17										3	1	4
	Gastritis	46				26	18	44	1	1	2						
	Ulcers	28				7	20	27		1	1						
Duodenum	Healthy	23	10	9	19										4		4
	Duodenitis	47				30	14	44	1	2	3						
	Ulcers	25				5	19	24		1	1						
Gallbladder	Healthy	24	9	11	20										3	1	4
	Gallstone Cholecystitis	17				13	2	15		2	2						
	Acute	16				3	12	15		1	1						
	Chronic	11				8	2	10		1	1						
Kidneys	Healthy	10	3	5	8										1	1	2
	Pyelonephritis	27				14	12	26	1	1							
	Nephrolithiasis	19				3	15	18	1	1							
Urinary Bladder	Healthy	10	4	4	8										1	1	2
	Cystitis	17				9	7	16	1	1							
TOTAL		516			175			288			17						36

Statistically significant difference between the total sum of true and false results: P<0.0001
Detection rate = 89.7% (86.8–92.0%); Sensitivity rate = 94.4% (91.2–96.5%); Specificity rate = 82.9% (77.2–87.4%); Predictive value rate (pos.) = 88.8% (85.0–91.8%) Predictive value rate (neg.) = 91.1% (86.2–94.4%)

TABLE 4.3 Comparison of Clinical Diagnoses and OED Results Obtained by Means of the Diagnotronics Device (0°–"Healthy," I°–"Within Normal Limits," II°–"Subacute," III°–"Acute") at the Department of Surgery, Helen Joseph Academic Hospital, Johannesburg

Organ	Clinical Diagnosis	No. of Subjects	True OED Results							False OED Results						
			Negative			Positive				Negative				Positive			
			0	I°	Together	II°	III°	Together	0	I°	Together	II°	III°	Together
Esophagus	Healthy	50	21	25	46							4		4
	Esophagitis	18				10	6	16		2	2			
	Cancer	16				1	14	15		1	1			
Stomach	Healthy	39	9	25	34							4	1	5
	Gastritis	29				19	5	24	2	3	5			
	Ulcers	11				1	9	10		1	1			
	Cancer	6				1	5	6			0			
Gallbladder	Healthy	66	17	42	59							6	1	7
	Gallstone	30				9	17	126		4	4			
	Cholecystitis													
	Acute	5					5	5						
	Chronic	4				2	1	3	1		1			
Pancreas	Healthy	51	8	33	41							8	2	10
	Pancreatitis													
	Acute	9				1	8	9			0			
	Chronic	12				7	3	10		2	2			
	Cancer	3				1	2	3			0			

4.3 CLINICAL ASSESSMENT OF OED ACCURACY

Organ	Condition										
Colon	Healthy	21	10	7	17					4	4
	Colitis	12				8	2	10	2	2	
	Cancer	11				2	7	9	2	2	
Kidneys	Healthy	108	33	63	96					10	2	12
	Pyelonephritis	14				11	2	13	1	1	
	Hydronephrosis	26				5	19	24	2	2	
Urinary Bladder	Healthy	72	30	33	63					8	1	9
	Cystitis	23				17	4	21	2	2	
	Cancer	21				4	15	19	2	2	
Prostate	Healthy	32	17	11	28					3	1	4
	BPH	11				10	0	10	1	1	
	Cancer	14				2	11	13	1	1	
TOTAL		714			384			246		29	55

Statistically significant difference between the total sum of true and false results: P<0.0001
Detection rate = 88.2% (85.6–90.5%); Sensitivity rate = 89.5% (85.2–92.8%); Specificity rate = 87.5% (84.0–90.4%); Predictive value rate (pos.) = 81.7% (76.9–85.9%); Predictive value rate (neg.) = 93.0% (90.1–95.2%)

of clinical diagnoses of the lungs and bronchi with OED results indicated that pneumonia usually manifests as "Acute" and bronchial asthma as "Subacute." The OED results for smokers were never "Healthy," but were at best "Within normal limits." Comparison of clinical diagnoses of the esophagus with OED results showed that esophagitis manifests most often as "Subacute," whereas cancer (advanced) produces the result "Acute." Comparison of clinical diagnoses of the stomach or duodenum with OED results showed that gastritis or duodenitis manifest most often as "Subacute," whereas ulcers and cancer (advanced) usually produce the result "Acute." According to the investigations, OED can detect even mild gastritis/duodenitis changes, which are not detectable with radiological methods. Comparison of clinical diagnoses of the gallbladder with OED results showed that acute cholecystitis manifests as "Acute," whereas chronic cholecystitis yields more distributed results. Asymptomatic gall stones produce the OED result "Subacute" or "Within normal limits," whereas symptomatic gall stones display "Acute." Comparison of clinical diagnoses of the pancreas with OED results showed that acute pancreatitis manifests mainly as "Acute," whereas chronic pancreatitis yields more distributed results. Comparison of clinical diagnoses of the colon with OED results showed that colitis manifests mainly as "Subacute," whereas cancer tends to yield "Acute." Comparison of clinical diagnoses of the kidneys with OED results showed that pyelonephritis manifests mainly as "Subacute," whereas hydronephrosis (various origins) tends to yield "Acute." Comparison of clinical diagnoses of the urinary bladder with OED results showed that cystitis manifests mainly as "Subacute," whereas cancer tends to yield "Acute." Comparison of clinical diagnoses of the prostate with OED results showed that benign prostate hypertrophy (BPH) manifests mainly as "Subacute," whereas cancer tends to yield "Acute."

Both studies confirmed that healthy organs usually display the OED result "Healthy" or "Within normal limits," whereas subacute pathology displays "Subacute" and acute pathology "Acute." The OED results were affected neither by the type nor the etiology of disease; that is, OED estimated the actual extent of the pathological process activity within particular organs, but it did not directly explain the cause of pathology.

It was observed that the OED results were not influenced by a patient's muscular tension, emotional state, skin humidity, or environmental temperature or by the procedure duration. The pressure of the measuring electrode had a limited influence (up to 5%) on the OED results and did not affect final diagnoses.

No side effects of the OED examinations were observed (60, 68).

4.4 DISCUSSION

For centuries, due to limited technical means, doctors could only make diagnoses "from outside." The OED opens a new chapter in the history of medicine by allowing direct access to the precise diagnostic information circulating in the body's own primary information system, which is the sensory nervous system. Various aspects of this new diagnostic approach are discussed in the following sections.

4.4.1 Innovative Aspects of OED

The sensory nervous system is created especially to detect any damage done to the body from both outside and inside and to send that information, at the earliest stage of pathology, to the

CNS, which controls potent self-defense mechanisms. Accessing this natural, firsthand source of accurate diagnostic information would open an entirely new era in medical diagnostics and even therapy, because early detection of pathology usually allows complete recovery. It appears that OED follows this medical logic, initiating a new generation of medical diagnostics that get access to the body's own information network in a noninvasive way in an attempt to obtain precise data about the actual state of the internal organs.

Diagnostic information coded in the frequency of the afferent nervous signals, sent from nociceptive receptors to the decision-making centers of the CNS, indicates the location of the pathology and the extent of the pathological process activity. Therefore, OED detects diseased internal organs/body parts ("regional diagnostics") and precisely estimates the activity of disease within these organs/body parts, but it does not directly explain the cause of pathology. In practice, this means that in the case of an OED result indicating pathology, conventional diagnostic methods should be used to estimate etiology and the kind of pathology. For the same reason, OED is unable to detect certain systemic diseases in their early stages (hypertension, diabetes mellitus, or HIV infection, for example), before the respective target organs are affected.

One can compare the OED to the diagnostics of modern cars; the "defectoscope" is connected to the car's computer terminal, and the list of faults is displayed on the screen.

4.4.2 Prospective Applications for OED in Contemporary Medicine

Clinical trials (60, 67, 68) confirmed that OED is a reliable bioelectronic method of noninvasive medical diagnostics, with high rates of sensitivity, specificity, and predictive values. The fact that the negative predictive values are higher than the positive predictive values suggests that OED may be relatively oversensitive. However, no clinical follow-up was done; OED could have detected pathology earlier than the comparative clinical methods. OED produces unequivocal diagnostic results immediately, with no need for any additional calculations. The use of optimal measuring parameters ensured accurate diagnostic results while avoiding any unpleasant sensations. Special attention should be paid to the ability of OED to investigate organs that are not easily accessible by means of standard diagnostic methods. Furthermore, OED makes a rapid assessment of all internal organs possible. The OED procedure is painless, easy to perform, quick, and cost-effective; therefore, the technology would be well suited for both first-line assessment and regular screening examinations for the early detection of any pathology, including breast, uterus, or prostate cancer.

This method not only detects diseased organs, but it also estimates the extent of the pathological process activity. OED readings can be compared to a laboratory ESR test. However, ESR only estimates the intensity of certain pathological processes in general (TB, for example), whereas OED can estimate the intensity of a pathological process in particular organs. The possibility of utilizing OED in monitoring the course of chronic diseases as well as for the early estimation of the efficacy of treatment has therefore become evident.

The sensory nervous system carries nociceptive signals only from the damaged tissues. This means that the OED can be very helpful in the proper diagnostics of various unclear medical problems. For instance, a benign breast tumor detected with mammography or an ultrasound examination will not change an OED result, but a malignant tumor will cause an "under the norm" OED reading. Also, nearly all adults show abnormal changes on spinal X-rays and scans, but this does not mean that nearly all adults suffer pain. The reverse is also true: Many people, especially younger ones, cannot even move because they have so much pain, but their X-ray and scan images look perfect. In all of these cases, OED can significantly contribute to the proper final diagnosis.

Pain is afferent impulses multiplied by central modulation (this can be influenced pharmacologically, psychologically, or surgically). The OED is able not only to trace the source of painful sensations but also to assess the frequency of afferent impulses sent from damaged tissues to the CNS. In this way, OED creates the possibility of a new objective pain assessment that can be widely used: for example, for disability grant evaluation purposes. If OED does not detect afferent impulses, there is no basis for a viscerogenic pain.

APs selected according to classical acupuncture rules for therapeutic stimulation usually display specific changes (see Section 2.5.4) in their electrical characteristics (65, 72, 75, 76). APs that do not correspond to diseased organs do not display these changes. This suggests the usefulness of OED in the scientific individual selection of optimal skin zones for reflexive therapies; in the case of the Diagnotronics device, only skin areas assessed as "Acute" or "Subacute" should be stimulated! In this way, OED may increase the efficacy of reflexive therapies and create the foundation for scientific acupuncture. Thanks to OED, the evidence-based map of auricular OPAs was created. Other body parts, such as feet or hands, can also be evaluated in the way described in Section 4.2 in order to prove the potential existence of OPAs on their surfaces (see Section 5.8.4.1).

4.4.3 Analysis of the Risk Associated with OED

A risk associated with this method is the possibility of misdiagnosis due to incorrect placement of the measuring electrode, similar to the risk of misplaced ECG or EEG electrodes. If, for example, the operator is intending to assess the condition of the lungs, and the measuring electrode is placed at the OPA corresponding to the heart, the result would be misinterpreted.

Various methods have been therefore implemented in the Diagnotronics device to minimize this risk. A high-resolution graphics display clearly indicates where the electrodes should be placed during each measurement. The software requires that each result be verified with a second measurement before the final diagnosis is specified. In addition, it is recommended that prospective operators should undergo training courses. The interpretation of OED results and the decision about whether to pursue further diagnostic/therapeutic procedures should be made by the doctor in charge. A false negative OED result in the case of a symptomatic patient would be corrected by other examinations. In the case of an asymptomatic person, a false negative OED result should not prevent such a person from attending comprehensive regular medical examinations, because in general there are no 100% accurate diagnostic methods in medicine.

OED will not replace existing diagnostic methods; instead, it provides additional information. An important benefit of the OED technology is that it can evaluate internal organs that could not otherwise be examined on a regular basis due to the cost and/or risk posed by existing techniques. Therefore, the relatively small risk (less than 10%, according to clinical trials) posed by a false negative OED result must be weighed against the probability that no examination would have taken place at all.

CHAPTER 5

Reflexive Physical Therapies

OUTLINE

5.1 General Remarks	75
5.2 Thermotherapy	77
5.2.1 Therapeutic Application of Heat	77
5.2.2 Therapeutic Application of Cold	78
5.3 Hydrotherapy	80
5.4 Ultrasound Therapy	80
5.5 Phototherapy	82
5.5.1 Infrared Radiation	82
5.5.2 Ultraviolet Radiation	83
5.5.3 Laser Stimulation	86
5.6 Electrotherapy	88
5.6.1 Transcutaneous Electrical Nerve Stimulation (TENS)	88
5.6.1.1 Electro-Sleep Therapy	98
5.6.2 Direct Electrical Nerve Stimulation by Use of Implantable Pulse Generators	99
5.6.3 High-Frequency Electromagnetic Fields	100
5.6.3.1 d'Arsonval's Currents	100
5.6.3.2 Shortwave Diathermy	101
5.6.3.3 Pulsed High-Frequency Magnetic Field Therapy	104
5.6.3.4 Microwave Diathermy	105
5.7 Magnetotherapy	106
5.8 Reflexive Mechanical Stimulation	109
5.8.1 Acupuncture ("Dry Needling")	109
5.8.1.1 History	110
5.8.1.2 Basic Principles of Modern Acupuncture	162
5.8.1.3 Chosen Examples of Acupuncture Therapy	167
5.8.2 Auriculotherapy (Acupuncture of the Ear Auricles)	214
5.8.3 Craniotherapy (Scalp Acupuncture)	217
5.8.4 Reflexive Massage	219
5.8.4.1 Reflexology (Reflexive Massage of Feet/Hands)	220
5.8.4.2 Tsubo Therapy	220
5.8.5 Cupping	220

5.1 GENERAL REMARKS

The human body is well equipped with various potent self-defense mechanisms and systems to preserve its homeostasis. For instance, it is by far the most advanced "pharmaceutical factory" and is able to synthesize any needed substance. Yet the immunological systems of people with AIDS do not react, leading them to die, simply because HIV is "smart" enough to kill the messengers. This example illustrates well the importance of the transfer of information in the human body. To make the full use of the body's powerful self-defense mechanisms and systems, the right signal must reach the right decision-making center, which does not always happen under all pathological circumstances.

Sometimes the mechanisms and systems that are supposed to protect the body's homeostasis actually aggravate the problem. Any serious injury or acute pathology, such as "acute abdomen," automatically causes as a firstline defense the generalized contraction of various muscles in the neighborhood of the problem area, including small muscles in the local arteries' walls, in order to diminish the extent of damage and prevent blood loss. In the case of something such as bronchial asthma, however, the bronchospasm, which is a result of the activation of the local nervous reflex arc by an inflammatory allergic reaction in the bronchial mucous membrane, can lead to fatal consequences. In the case of ureterolithiasis, in which the stone traversing the ureter hurts the ureter mucosa from inside and in this way triggers the local reflex arc, it will eventually result in the strong contraction of the ureteric muscles and the "imprisonment" of the stone. The stone then firmly blocks the urine flow, and the kidney swells like a balloon (hydronephrosis), causing extremely painful "renal colic." A vicious circle is thus created: the stronger the urine pressure on the stone, the stronger the muscle spasm and the urine blockage. There are many similar vicious circles taking place in the human body due to defects in the biological transfer of information.

By using various forms of physical energy to stimulate skin nervous receptors and in this way control the flow of information in the nervous system, the body's primary computing network, reflexive physical medicine reprograms self-defense mechanisms and systems according to actual needs (see Chapter 3). Because reflexive therapies work directly on the peripheral nervous system, the best therapeutic effects are observed at these lines of the body's self-defense, which are related to this part of the nervous system; these self-defense markers include pain, muscle spasms, and local blood supply. Pain = afferent impulses × central modulation (this can be influenced pharmacologically, psychologically, or surgically); reflexive physical medicine blocks afferent impulses, removing the basis for a viscerogenic pain. Reflexive therapies always block the pain in the same manner, no matter what kind of pathology causes it; in this way, they work as a regional analgesia. Hormonal and humoral changes come as a secondary response to reflexive treatments (see Chapter 3). Therefore, reflexive physical medicine should not be the first choice when dealing with purely endocrynological, immunological, oncological, infectious, or psychiatric problems. To date, reflexive therapies have been grossly underutilized, mainly due to the lack of clear understanding of their mode of action. After all, biocybernetics and other computer sciences, which can explain these mechanisms in the best way, were developed only recently. Nevertheless, because it is clinically effective, natural, and cost effective, reflexive physical medicine carries the potential of better healthcare at a much lower cost.

All reflexive therapies, irrespective of their therapeutic efficacy, work the same way (see Chapter 3): They stimulate dermal nervous receptors, which generate afferent signals. These signals, which consist of chains of nervous active potentials, do not differ from each other depending on the nature of the stimulus, but the frequency of active potentials is proportional to the intensity of stimulation (including the amplitude of a stimulus). Therefore, certain reflexive therapies can be more effective than others; but in general, they all produce very similar final clinical effects: pain relief, relaxation of muscle spasms, improved local blood circulation, and, therefore, reduced inflammation and edema as well as increased metabolic rate. Indicators of increased metabolic rate—that is, increased local synthesis of adenosine triphosphate (ATP), ribonucleic acid (RNA), hydroxyprolin (important in collagen synthesis), prostaglandins, and other proteins—often reflect a restoration process of the tissue damages caused by applied physical factors and should not be wrongly interpreted as the indicators of the therapeutic efficacy of the physical stimulation. In fact, one can expect much higher indicators of increased local metabolism after the application of more harmful stimuli such as laser, DC, or ultraviolet radiation compared to the use of, for example, reflexive massage, AC, or "ice packs." However, this does not mean that these more harmful therapies are more clinically effective.

Efficacy of reflexive physical medicine depends mainly on the choice of skin areas selected for therapeutic stimulation; all of these areas should display tenderness (practical but subjective criterion) and a high degree of rectification/increased impedance phenomenon (objective scientific criterion) (see Sections 3.3 and 4.4.2). Reflexive physical medicine efficacy depends also on the amplitude and intensity of the stimulus (26, 27): generally, the higher the better, as long as it is safe and does not induce unpleasant sensations. We always have to remember that a stimulus in high-energy form, for example, laser knife, electrocoagulation, radiofrequency rhizotomy, intensive shortwave/microwave diathermy, or extreme heat/cold, will destroy/or damage the tissue (stimulating nociceptive receptors), whereas in the therapeutic low-energy form, it will only stimulate non-nociceptive nervous receptors. Therefore, the stimulus parameters must be well balanced to ensure safety, prevent unwanted side effects, and yet preserve the maximal therapeutic efficacy. The kind of stimulus is less important; it is good to combine various stimuli in order to prevent the adaptation of nervous receptors.

Reflexive therapies in which a constant stimulus is used, such as constant heating or cooling, DC stimulation, static magnetic fields, continuous laser stimulation, or classic acupuncture (without needle manipulation), are therapeutically less effective than therapies using the same stimuli in dynamic (intermittent/modulated) form, such as alternating heating and cooling, modulated AC stimulation, pulsated electromagnetic fields, pulsated laser stimulation, or electroacupuncture (19, 26, 27, 39). This is because modulated stimulation prevents the adaptation of nervous receptors. Dynamic stimulations, especially with high amplitude and intensity (within safe limits) induce higher frequencies of afferent nervous impulses. The frequency range of these dynamic therapies is generally 1–200 Hz, which coincides with the frequency range that is best detectable by dermal nervous receptors. Higher stimulating frequencies do not increase nervous activity (82, 86). Interestingly, pulsed (intermittent) stimulation generally applies a significantly lower dosage of energy to the body compared to constant stimulation; this phenomenon supports the reflexive explanation of the physical medicine mode of action.

All of the previously discussed facts concerning the efficacy of reflexive physical medicine explain why certain methods, such as acupuncture and other related techniques, TENS, extreme cryotherapy (−70 or, better, −160 degrees Celsius) are clinically more effective than lasertherapy, ultrasoundtherapy, shortwave/microwave diathermy, magnetic field therapy, hydrotherapy, "hot packs," etc. Nevertheless, as noted earlier, it is good to mix these methods altogether.

When it comes to the frequency and duration of therapeutic procedures, reflexive therapies can be used intensively, over a short period of time, for emergency purposes (such as status asthmaticus, renal colic, atonic uterus, esophagospasm, severe migraine, or postoperative pains). However, chronic conditions require at least several treatments on a regular basis to ensure a lasting improvement. Our unpublished radioisotopic research showed that significantly improved blood supply to the treated body region lasts only up to 24 hours after initial treatments; therefore, therapeutic sessions should be repeated at least every second day. Usually, there is no visible improvement during the first several sessions; even if there is any, it will not be lasting. Then, during the so-called transitory period of treatment, "good" days alternate with "bad" days; the duration of this treatment period varies individually, depending on the problem treated. Finally, "good" days prevail and "bad" days subside. Interestingly, about 20% of all the patients treated with reflexive physical medicine report aggravation of their symptoms after the initial one to five treatments; in fact, this is a very positive sign, because usually all of these patients eventually show a lasting clinical improvement. The usual procedure duration depends on the type of treatment; for example, it cannot be too long in the case of cryotherapy (−160 degrees Celsius) or moxa, but it generally oscillates between 30 and 60 minutes. Reflexive therapies break the vicious *pain–muscle spasm–pain* circle, and it takes time for it to restart; this is why the relief lasts usually much longer than the therapeutic session itself.

In the following sections, various methods of reflexive physical medicine will be discussed; the most attention will be paid to the modern and clinically most effective ones rather than those of only historical importance.

5.2 THERMOTHERAPY

The therapeutic application of heat and cold to the human body has a long tradition. Certain classical methods, such as "dry Roman baths," saunas, or ice-cold baths (often replaced these days by cryotherapeutic chambers), are still in use but for recreational purposes rather than for typical medical indications. In this chapter, therefore, we will concentrate on specific reflexive thermal therapies.

5.2.1 Therapeutic Application of Heat

Heat-based therapies belong to a moderately effective group of methods of physical medicine (see Figure 3.4) because their stimulus amplitude is limited: temperatures higher than 50 degrees Celsius pose the risk of burns.

Many specialist centers of physical therapy still use paraffin wrapping or local paraffin baths. Good calorific capacity and poor thermal conductivity make paraffin useful for therapeutic purposes due to slow thermal voidance. A standard treatment includes an application of hot paraffin (40–50 degrees Celsius) directly on the affected body parts for 30–60 minutes. This kind of thermotherapy is especially useful for arthritis and soft tissue inflammation.

Much more popular, especially when it comes to home treatment, are "hot packs." These are usually small, portable bags filled with various, usually natural substances characterized by a good calorific capacity that can be easily warmed up (for example, in the microwave) and then applied to the affected body parts. An older version of a hot pack is represented by a "thermophor," a rubber/plastic bag filled with hot water. Hot packs are mostly used for back pain, arthritis, and soft tissue inflammation.

Various electrical heating devices, including "electric pillows" and "electric sheets," have become more popular in recent times. One advantage these devices have over hot packs is continuous temperature control (at a maximum 50 degrees Celsius) during the entire treatment, which can therefore be prolonged. The range of therapeutic indications is similar to that of hot packs.

In the Far East, moxibustion is still very popular. This specific type of thermotherapy consists of making a small, moistened cone (moxa) of powdered leaves of mugwart or warmwood (Artemisia species), applying it to the skin (APs), igniting it, and then crushing it into the blister so formed. Other substances can be used for the moxa as well. Small, slowly burning "cigars" can also be fixed on acupuncture needles inserted into the skin or placed on special leaves that act as a thermal barrier to prevent skin burns. On most occasions, however, after completion of the course of these treatments, permanent scars can be seen on the skin. The usual duration of the heat application is a few minutes per each stimulated skin area. The course of treatments includes several sessions on a regular basis depending on the kind of problem. Moxibution is among the more effective and versatile methods of physical medicine as long the stimulated skin areas are properly chosen (see Section 5.8.1.2.1).

It seems that certain traditional healing techniques based on placing the hands over the affected body parts (Reiki, Bioenergotherapy, etc.) can also use the healer's body heat as a thermal stimulus. Because this type of stimulation is very soft, these "bioenergetic" therapies have a limited level of clinical efficacy.

There is no direct contraindication for the use of heat-based therapies. However, moxa treatments applied to the lower back, legs, and abdomen should be avoided for women in pregnancies with a high risk of miscarriage.

5.2.2 Therapeutic Application of Cold

Cryotherapy is among the most effective methods of physical medicine (see Figure 3.4) because its stimulus amplitude can be very high. However, it only poses the risk of burns if the time of application is too long. Rapid refrigeration takes a limited amount of warmth away from the superficial tissues, yet it strongly stimulates the dermal nervous receptors: a clear indication of the cryotherapy reflexive mode of action! Even better is a combined thermotherapy: alternating applications of heat and cold (1–5 minutes each) makes the stimulus amplitude even higher.

Some traditional physical therapy centers still use ice wrapping: sheets are refrigerated to a requested temperature, or a rubber/plastic bag (thermophor) is filled with crushed ice. What are far more popular these days, however, especially for home treatment, are ice packs. These are usually small, portable plastic bags, filled with a special gel, which can be chilled in the refrigerator and then applied to the affected body parts.

For small surgical procedures or sports injuries, local application of liquid ethyl chloride is particularly useful; it effectively freezes the respective body surfaces and, due to the strong stimulation of skin non-nociceptive receptors, blocks pain signals originating from the area.

The therapeutically most effective approach, however, is so-called extreme cryotherapy involving the vapors of liquid nitrogen (Figure 5.1) or liquid carbon dioxide, which produce thermal stimuli at, respectively, −160 and −70 degrees Celsius. This type of therapy was introduced in 1979 by the Japanese doctor T. Yamauchi. Cryotherapeutic devices of this kind include a reservoir for a liquid gas, a heater to produce a vapor (liquid nitrogen) or a regulating valve (liquid carbon dioxide), and a pipe with a nozzle to apply the cold vapor to affected body parts. During the treatment, the cold vapor should not be applied continuously to the same skin spot but rather should be applied all around the stimulated skin area to avoid frostbite. Such an intermittent application can also be therapeutically more effective than constant stimulus, which leads to nerve adaptation. The usual duration of the cryotherapeutic procedure is up to one minute per stimulated skin area; it must be stopped once the whitish coloration of frozen skin appears. If indicated, the treatment can be repeated up to three times per day. It is good for patients to start intensive therapeutic exercises immediately after cryotherapy procedures; they will be able to perform even those exercises that otherwise would be very difficult for them. Therapeutic application of heat (3–5 minutes) prior to cryotherapy will enhance the final therapeutic effect (26, 27).

There is a wide range of clinical indications for cryotherapy that includes rheumatoid, psoriatic, and osteoarthritis, ankylosing spondylitis, acute gout, persistent back pains, neuralgias, neuropathies, reflexive sympathetic dystrophy, phantom pains, allergic rhinitis and chronic sinusitis, bronchial asthma, renal colic, soft tissue injuries, posttraumatic edema (such as after bone fracture), lymphatic edema post mastectomy, and, especially, fresh burns. So-called extreme cryotherapy chambers, which can accommodate several people at the same time with only their protruding body parts (fingers, ear auricles, nose, penis, and testicles) covered with gloves, headbands, nose masks and "G-strings" to avoid the risk of frostbite, are used mainly for recreational and sports

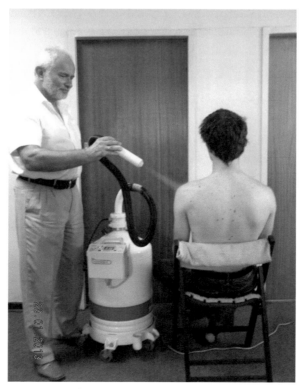

FIGURE 5.1 Extreme cryotherapy procedure with vapors of liquid nitrogen (−160 degrees Celsius).

medicine purposes. However, this whole-body cryotherapy (1–3 minutes) can be very useful for the treatment of widespread pathologies such as polyarthritis.

Cryotherapy should not be used in case of cryoglobulinemia and should be carefully considered in case of Raynaud syndrome. In a pregnancy with a high risk of miscarriage, it should not be applied locally to the lower back, legs, and abdomen.

5.3 HYDROTHERAPY

Hydrotherapy belongs to the classical methods of physical medicine. It combines together thermotherapy and mechanical stimulation. The water displays important physical features: its thermal conductivity is about 25 times better and its specific heat a few thousand times higher than that of air. Another important factor is that the body dipped in hot water loses its ability to cool by perspiration. Also, water vapor can be used for physical medicine purposes, although Turkish steam baths (which were historically important) play a more recreational role these days.

The most important contemporary form of hydrotherapy is the therapeutic shower. There are cold (10–15 degrees Celsius), warm (25–30 degrees Celsius), hot (38–42 degrees Celsius), and mixed temperature showers. There are also low (below 150 kPa), medium (150–200 kPa) and high (200–400 kPa) pressure showers. The shower can be mobile—with the stream of water/or steam applied all around the stimulated skin area—or a so-called whip shower—with a high-pressure, intermittent stream of water/or steam. The very effective "Scottish" shower combines a high-pressure intermittent stream of water with alternating water temperature; in this way, it applies the highest amplitudes of both stimuli. The usual duration of therapeutic showers is 2–3 minutes. The indications include neuroses, neuralgias, various skeletal disorders and diseases, chronic respiratory tract diseases, and disorders of peripheral blood circulation. Therapeutic showers should not be used in patients with high blood pressure, unstable coronary heart disease, or heart failure. They should be carefully considered in cases of epilepsy and in pregnancies with a high risk of miscarriage.

Another hydrotherapy form, which continues to gain popularity, is the Jacuzzi. This is a bathtub equipped with underwater jets. Hot water massage, usually with the addition of air bubbles, is useful for spastic paralyses, back pains, neuralgias, various kinds of arthritis, soft tissue injuries, impotence, and even insomnia. For patients with cardiac problems, the water should not be too hot.

5.4 ULTRASOUND THERAPY

Ultrasounds are mechanical vibrations characterized by a frequency beyond the hearing range of the human ear. These vibrations are transferred to the neighboring molecules, and in this way an ultrasound wave is created. When an ultrasound wave approaches a medium of different density, it is deflected entirely or partially; this phenomenon is widely utilized in ultrasound diagnostics. The intensity of the ultrasound wave decreases proportionally to the distance from the source of vibrations due to energy absorption by the environment. The absorption depends on the ultrasound frequency and the environment's characteristics. The highest absorption ability is displayed by gases, a lower absorption by fluids, and the lowest by springy solid bodies

FIGURE 5.2 Ultrasound therapy by means of a mobile applicator.

such as metals, which are good vibration conductors. Elastic solid bodies such as rubber or cork significantly absorb sounds and therefore can be used as acoustic insulators.

Various human tissues are characterized by various degrees of ultrasound absorption. A high degree of absorption is displayed by the nervous tissue, a lower absorption degree by muscles, and the lowest by the adipose tissue. Shorter ultrasound waves with a higher frequency are absorbed more superficially, whereas longer waves are absorbed more deeply.

Ultrasoundtherapy displays a good stimulating effect on nervous receptors, because it combines direct mechanical stimulation—so-called micromassage—with thermal stimulation; as was noted previously, nervous tissue is a very good absorber of ultrasound energy.

The therapeutic range of contemporary ultrasoundtherapy varies from 0.1 to 30 W/cm^2, but dosages higher than 2 W/cm^2 are very seldom applied. Weaker dosages should be used for areas with limited soft tissue such as the face. Higher ultrasound frequencies, 2400 kHz or more, are more suitable for reflexive therapy purposes than the widely used therapeutic frequency of 800 kHz, because higher-frequency ultrasound waves are absorbed more superficially. Modulated pulse emission is preferred (compared to constant emission), because it will prevent the adaptation of nervous receptors. The procedure duration depends on the method of application; for a stationary applicator (a high concentration of ultrasound waves), (Figure 5.2) it is usually 1–3 minutes. For a mobile applicator (ultrasounds applied all around the stimulated skin area), it is usually 3–10 minutes. Ultrasounds are usually applied directly to the affected body parts as well as to "trigger points" via liquid paraffin or special gels to insure a good contact between the applicator and the skin. The treatments can be done on a daily basis for more acute problems and every second day in case of chronic pathologies.

The example indications for ultrasoundtherapy and the usual parameters (using a mobile applicator) follow:

- Upper back pain: 0.5–0.8 W/cm^2 up to 10 minutes altogether; 12–20 treatments per course
- Dorsal back pain: 0.5–0.8 W/cm^2 up to 10 minutes altogether; 12–20 treatments per course
- Lower back pain: 0.8–1.2 W/cm^2 up to 10 minutes altogether; 12–20 treatments per course
- Hip osteoarthritis: 0.5–1.0 W/cm^2 up to 10 minutes altogether; 12–20 treatments per course
- Knee osteoarthritis: 0.3–0.8 W/cm^2 up to 7 minutes altogether; 12–20 treatments per course

- Shoulder osteoarthritis: 0.4–0.8 W/cm² up to 10 minutes altogether; 12–20 treatments per course
- Elbow osteoarthritis: 0.3–0.5 W/cm² up to 10 minutes altogether; 12–20 treatments per course
- Hands and feet osteoarthritis: 0.3–0.8 W/cm² up to 8 minutes per each hand/foot; 12-20 treatments per course
- Trigeminal neuralgia: 0.3–0.5 W/cm² up to 8 minutes altogether; 12–20 treatments per course
- Chronic sinusitis: 0.3–0.5 W/cm² up to 5 minutes per sinus; 12–20 treatments per course
- Bronchial asthma: 0.5–0.8 W/cm² up to 3 minutes per each "trigger point" on the posterior and frontal chest; 12–20 treatments per course
- Tendonitis: 0.5–0.8 W/cm² up to 8 minutes altogether; 12–20 treatments per course
- Phantom pains: local and respective spinal segment stimulation: 0.5–0.8 W/cm² up to 10 minutes per each area; 12–30 treatments per course

Contraindications for ultrasoundtherapy include active tuberculosis and malignant tumors. Ultrasoundtherapy should not be applied to body areas that include metal implants. In pregnancies with a high risk of miscarriage, it should not be applied to the lower back, legs, and abdomen.

5.5 PHOTOTHERAPY

Historically, the oldest form of phototherapy is so-called heliotherapy: the therapeutic utilization of solar radiation. Solar radiation is the main source of vital energy on Earth. A total of 59–65% consists of infrared radiation (IR), 33–40% visible radiation, and 1–2% ultraviolet radiation (UV). Solar radiation (especially IR) is partially absorbed by the layers of the atmosphere, in particular steam, carbon dioxide, and ozone. This "protective coat" around the Earth also prevents the escape of the Earth's own IR into the cosmos. The intensity of solar radiation also depends on the seasons of the year and the time of the day, which determine its angle of incidence. The therapeutic effect of heliotherapy is a result of the application of both IR and UV to the skin.

5.5.1 Infrared Radiation

IR is an invisible electromagnetic radiation. In the electromagnetic radiation spectrum, it is located between the ranges of a visible red radiation and microwaves. For therapeutic purposes, the IR wavelength range of 770–15 000 nm is used. Depending on the wavelength, the IR can be divided as follows:

- Shortwave IR: wavelength is from 770–1500 nm
- Mediumwave IR: the wavelength is from 1500–4000 nm
- Longwave IR: the wavelength is from 4000–15 000 nm

About 30 % of IR applied to the skin is reflected; the rest penetrates inside the skin. The penetration ability depends on the wavelength. Shortwave IR penetrates about 30 mm inside the tissues, but it is absorbed mainly at a depth of up to 10 mm. Longwave IR penetrates only 0.5–3 mm deep. Therefore, any therapeutic influence of IR on tissues located deeper under the skin can be best explained in the reflexive way.

IR application to the skin causes the dilatation of local blood vessels, leading to the visibly reddish coloration of the stimulated skin area. This thermal erythema occurs during IR application

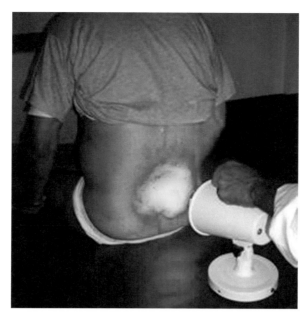

FIGURE 5.3 Local IR therapy.

and intensifies depending on the duration and intensity of the IR application. The reddish coloration of the skin is inconsistent (patches) due to the dilatation of the more deeply located blood vessels. Thermal erythema subsides soon after the IR application is finished; the duration of this type of erythema depends on the dosage of IR. However, local nerve damage can weaken or even prevent the appearance of erythema; this phenomenon directly indicates the leading role of the reflexive mechanism in IR therapy.

There are various IR therapeutic devices available on the market. Most of them, with a single source of IR, are used for local treatment of particular body parts (Figure 5.3). However, there are also special chambers, equipped with multiple IR sources, to treat vast areas of the body; these are used mainly for recreational purposes. The usual duration of the IR treatments is 20–30 minutes; they can be repeated 2–3 times per day. The distance between the source of the IR and the skin should insure maximal application of the heat yet prevent potential burns (a maximum temperature of 50 degrees Celsius). The staff supervising the IR therapy must wear sunglasses, because IR can contribute to the creation of cataracts.

IR therapy belongs to the category of moderately effective methods of physical medicine, similar to thermotherapy. It is used mainly for soft tissue inflammation, arthritis, back pains, wound/ or skin ulceration healing, and the treatment of skin damage after radiotherapy.

IR therapy is traditionally contraindicated in active tuberculosis. IR chambers should not be used in the case of heart failure.

5.5.2 Ultraviolet Radiation

UV is an invisible electromagnetic radiation characterized by the wavelength ranging from 100–400 nm. In the electromagnetic radiation spectrum, it is located between the ranges of a visible violet radiation and so-called soft X-rays. For therapeutic purposes, the UV wavelength range of 200–400 nm is used. Depending on its biological impact, the UV can be divided as follows:

- Range A: wavelength from 400–315 nm
- Range B: wavelength from 315–280 nm
- Range C: wavelength from 280–200 nm

The Schumann's UV of the wavelength ranging from 100–200 nm has no practical medical importance, because it is almost entirely absorbed by the air and the water vapor.

Various substances show various degrees of UV absorption. Quartz, which passes the radiation of the wavelength longer than 180 nm well, is widely used in UV lamps. Standard window glass, however, passes only the radiation of wavelengths longer than 320 nm. Chance–Crookes's glass absorbs the UV entirely and therefore is widely used in protective sunglasses.

UV is well absorbed by human skin, so it can penetrate inside the human body only up to 2 mm deep. Therefore, any therapeutic influence of the UV on the tissues located deeper under the skin can be best explained in the reflexive way. We also have to remember that part of the applied UV is reflected from the skin; this depends on the application angle, the condition of the skin, and the UV wavelength.

The UV causes so-called photochemical reactions: photosynthesis, photolysis, and photoisomerization. Photochemical reactions are responsible for, among other things, the creation of photochemical erythema in the skin, pigment creation, and vitamin D production. The UV bactericidal effect (particularly at the 250–270 nm wavelength) is also based on the photochemical reactions that lead to the structural changes in bacterial proteins and to blockage of the vital processes.

The term photochemical erythema describes the reddish skin coloration that is due to the dilatation of local blood vessels. The photochemical erythema's intensity depends on the UV wavelength, the intensity of the UV emission, the duration of the UV application, the distance between the source of radiation and the skin, skin sensitivity (especially the size of the epidermis), general carnation—blondes are more sensitive than brunettes—and age—children are more sensitive than adults.

There is a characteristic evolution of the photochemical erythema that includes the following phases:

- Latent period of the duration from one to six hours. Absorption of the UV by the epidermis cell proteins leads to their denaturation and cell damage. From damaged cells histamine and other vasodilating substances are released, causing dilation and increased permeability of the skin capillary vessels.
- Intensification period covers the time between appearances of the first symptoms of vasodilation and the maximal erythema's intensity, which usually takes place within six to 24 hours after exposure to the UV. During this period of time, a skin edema can appear, sometimes with water blisters between the skin layers. Excessively strong UV dosages can lead to irreversible damage and necrosis of the epidermic cells.
- Disappearance period of the duration from some hours to some days, depending on the UV dosage absorbed. As a result of the photochemical erythema, an epidermal thickening, scaling, and brownish skin discoloration (due to pigment accumulation) can be observed.

Contrary to the thermal erythema caused by the IR, the photochemical erythema includes a latent period, is solid, and is limited strictly to the exposed skin area. The IR application to skin area with the photochemical erythema can weaken the intensity and accelerate the disappearance of a photochemical erythema. Also, local nerve damage can weaken or even prevent the appearance of a photochemical erythema that indicates an important role of the reflexive mechanism in UV therapy. Certain chemical substances can significantly increase body sensitivity to UV, including coal tar, sulphonamides, tetracyclines, chlorpropamid, tolbutamol, promethazinum, diazepam, and salicylamids. So-called photodynamic agents, especially psoralens, are used to increase the efficacy of UV therapy. Increased sensitivity to UV can also be seen in certain diseases, including lupus erythematosus, porphyria, dermatomyositis, and xeroderma pigmentosum.

The UV application improves local blood supply and stimulates a higher metabolic rate; therefore, the skin becomes more elastic, looks younger, and is more resistant to infections. For these reasons, UV is widely used for cosmetic purposes (in sunbeds, for example). For the same reasons, UV therapy is also successfully used in the management of wounds, bedsores, and chronic skin ulcerations. The wavelength of more than 280 nm is usually used for these purposes, because wavelengths shorter than this cause epidermic damage. The A range of UV is widely used in the photochemotherapy ultraviolet A (PUVA), which combines the UV (especially wavelengths of 360–365 nm) with photodynamic agents and which is a therapy of choice in psoriasis and other skin diseases. The so-called selective UV-phototherapy (SUP) uses the wavelength from 300–340 nm and does not require photodynamic agents.

Prophylactic application of UV is commonly used in vitamin D shortages, especially as a prevention of rachitis. Special bactericidal UV lamps are commonly used in hospitals, health centers, and laboratories. However, physical medicine mainly applies UV in skin diseases (see previous note; also, acne vulgaris), alopecia areata, soft tissue inflammations, arthritis, neuralgias, bronchial asthma, nose and throat diseases (using special applicators), and even endocrine hypofunctions (thyroid and ovaries).

There are various kinds of UV generators available on the market. They can be portable for local use (Figure 5.4), but there are also special chambers (sunbeds, for example) for general

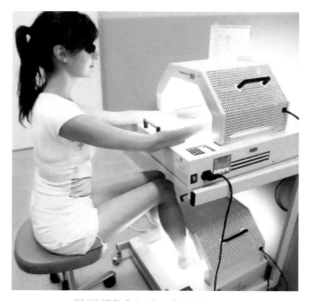

FIGURE 5.4 Local UV therapy.

FIGURE 5.5 UV general application by means of a sunbed.

applications (Figure 5.5). Depending on their technical parameters and special filters, they offer a variety of therapeutic options. Because UV overdose can lead to burns, tissue necrosis, and other serious complications, it is of utmost importance to follow the individual instruction manuals strictly, especially with regard to the range of applied wavelengths, the intensity of UV emission (including the distance from the UV source), and the treatment duration. Both the patient and the therapist must protect their eyes with special glasses.

UV therapy is generally contraindicated in cases of malignant tumors and also traditionally in active pulmonary TB as well as epilepsy.

5.5.3 Laser Stimulation

Laser therapy is one of the newest methods of physical medicine, yet it is surrounded by many myths and legends. The word "laser" is an abbreviation of the term "light amplification by stimulated emission of radiation" and is also used with regard to the respective devices: quantum amplifiers. In general, laser therapy belongs to the group of versatile but rather "soft" methods of physical medicine, because in low-energy form it offers only a mild stimulation of the dermal nervous receptors, and the use of higher energies could lead to the serious tissue damages. Nevertheless, because the stimulus parameters can be fully controlled, laser therapy is both convenient and safe. (Figure 5.6).

Laser radiation displays the following characteristics that distinguish it from conventional light radiation caused by a spontaneous emission:

- Cohesion: the laser lightwave consists of long and identical wave sequences capable of interference.
- Monochrome: the laser radiation is of an almost identical wavelength.
- Collimation: there is a very small divergence angle, up to 1 second.

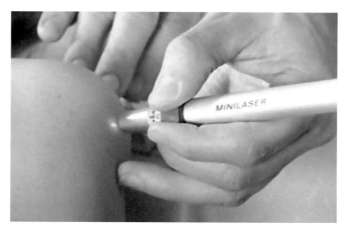

FIGURE 5.6 Laser therapy.

There are the following criteria for the laser classifications:

- Length of the emitted wave (UV, visibility range, IR)
- Kind of active medium (solid, liquid, gas, semiconductor)
- Work regimen (pulsating, continuous)
- Power: medical lasers can be qualified as high-power "surgical" lasers (10–100 W), mid-energetic
- "Photodynamic therapy" lasers (1–10 W) and low-power "biostimulating" lasers (10–400 mW)

Physical medicine usually uses "soft lasers" (power below 500 mW) for the so-called low-level laser therapy (LLLT). However, in the case of the pulsating mode of laser stimulation by means of very short pulses, such as 200 ns, the peak power can reach several watts; the energy applied to the body is still minimal, especially when the frequency is not higher than an optimal range 2–200 Hz (see Section 5.1). In this category, the most popular are semiconductor lasers "Ga-As" (wavelength 635–980 nm) and gas lasers "He-Ne," which emit visible red radiation (wavelength 632 nm). In general, shorter waves penetrate tissues no deeper than 15 mm and longer waves up to 30–50 mm.

The laser radiation applicator should be placed directly on the skin in order to avoid any energy losses due to reflection. However, in case of skin ulcerations or burns the applicator should not be touching the skin, yet the distance should be minimal. The laser radiation can be applied to particular skin spots, such as OPAs/APs, or it can be spread over the larger area of the skin; this so-called scanning can be done automatically by preprogrammed modern laser devices. The usual time of laser stimulation varies from one minute (skin spots) to 20 minutes; treatments can be done on a daily basis for up to 20 treatments (or more) per series. Application of the LLLT can increase the tissue temperature only up to 0.5 degree Celsius; it is not a thermal effect but clearly a reflexive mechanism responsible for the LLLT therapeutic results (see Section 3.3 in Chapter 3).

There is a wide range of indications for laser therapy, which includes following conditions:

- Wounds and chronic skin ulcerations, including bedsores: recommended wavelength 632–680 mm

- Slow healing of bone fractures: recommended wavelength 830–980 nm
- Soft tissue inflammation: recommended wavelength 635–980 nm
- Arthritis: recommended wavelength 830–980 nm
- Tennis elbow: recommended wavelength 830–980 nm
- Dupuytren's contracture; recommended wavelength 830–980 nm
- Back pains: recommended wavelength 830–980 nm
- Peripheral neuralgias, especially post-herpetic: recommended wavelength 830–980 nm
- Neuropathies, including diabetic: recommended wavelength 830–980 nm
- Acne vulgaris: recommended wavelength 632–680 nm

To insure the best results, all treatments should include both local laser therapy (scanning) and stimulation of the respective OPAs/APs (see Section 5.8.1.2.1). Precise localization of the right skin areas is of the utmost importance, because it determines the efficacy of the treatment; all stimulated skin areas should display tenderness (practical but subjective criterion) and a high degree of rectification/increased impedance phenomenon (scientific criterion) (see Sections 3.3 and 4.4.2). Because the laser stimulation of OPAs/APs can replace, at least to a certain extent, acupuncture needles, indications for laser therapy also include all those for acupuncture (see Section 5.8.1.3). Laser radiation can cause serious retinal damage, and so both the patient and the therapist must protect their eyes with special glasses which are different for different laser wavelengths.

The only contraindication for the use of LLLT is a malignant pathology; laser therapy can theoretically stimulate the growth of a neoplasm.

5.6 ELECTROTHERAPY

After acupuncture, electrotherapy is another method of physical medicine most surrounded by various myths, legends, and prejudices, especially when it comes to the explanation of its mode of action. Foundations for the therapeutic utilization of electrical current were created in the 18th century by the famous Luigi Galvani, who electrically induced spasms in frogs' muscles. Contemporary rehabilitation still widely uses electrical muscle stimulation for the treatment of muscle paralyses, correction of spinal scoliosis, and even slimming therapy. Cardiologists commonly use defibrillation, neurologists use rhizotomy, and surgeons use electrocoagulation; iontophoresis is used to enhance drug absorption through the skin. However, physical medicine uses electrostimulation (direct or via electromagnetic energy) for reflexive purposes. Because electrical stimulus parameters can be fully controlled, this kind of therapy is not only safe but also convenient and effective.

5.6.1 Transcutaneous Electrical Nerve Stimulation (TENS)

Any application of electrical energy to the skin causes stimulation of the dermal nervous receptors and therefore can be called transcutaneous electrical nerve stimulation (TENS). Historically, depending on the level of technical development, various kinds of therapeutic electrostimulation have been used. The initial use of DC, for so-called galvanization or hydro-electrical baths (four chambers, etc.), carried the risk of severe tissue damage when higher intensity was used (electrocoagulation) and displayed minimal stimulating effects with low values of voltage/

intensity; according to the Du Bois-Reymond's law, it is not an electrical current itself causing a stimulus, but sufficiently quick changes in its intensity. Therefore, there is no place for DC stimulation in modern electrotherapy. For the same reason, traditional unipolar pulsated currents with relatively long impulse duration, the so-called Traebert's current ("Ultra Reiz") with square impulses (2 ms width) and fixed frequency (about 140 Hz) or Bernard's currents ("diadynamic") with various combinations of semisinusoidal impulses (10 ms width) and frequencies of 50 or 100 Hz, are these days of only historical importance. Slightly better stimulating characteristics are shown in the classical Faradic current, which consists of "modulated" chains of asymmetrical impulses with various shapes and values of width and amplitude. However, the impulse durations also oscillate between 1 and 3 ms: this is more useful for muscle electrostimulation, but prevents obtaining higher impulse amplitudes, which are required for the maximal nerve stimulation.

According to the convergence modulation theory (see Section 3.2), the most effective electrotherapeutic current would be that which comprises the most "irritating" parameters: bipolar impulses with possibly high amplitudes and therefore short durations, low frequency, and modulation of all parameters. Symmetrical bipolar impulses not only ensure maximal amplitude of the stimulus but also prevent any possibility of skin burns under the electrodes (electrocoagulation, which depending on the current intensity and procedure duration always occurs whenever dealing with unbalanced electrical pulses). In order to keep the stimulation below the pain threshold, the use of high amplitudes of impulses requires adequately short durations of impulses to ensure an application of low dosage of energy to the skin. Low-frequency stimulations are better detected by nervous receptors (see Section 5.1); the use of so-called medium-frequency currents (1000–10 000 Hz) will not increase nervous activity (82, 86). Modulation of all electrical parameters prevents the adaptation of the nervous system. Interestingly, modern electrotherapeutic devices built on the basis of clinical experience are usually characterized by output parameters compatible with optimal ones proposed on the basis of the convergence modulation theory: wave—bipolar impulse, maximal impulse amplitude—at least 150 V, impulse width—from 50 to 300 μs, frequency—from 3 to 100 Hz (usually below 20 Hz), modulation—all parameters (Figure 5.7). There is in general minimal energy applied to tissues during the TENS treatment, which clearly indicates a reflexive mode of action of this kind of therapy.

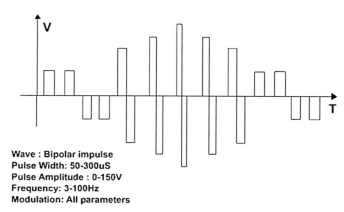

FIGURE 5.7 Standard electrotherapeutic parameters for modern analgesic electrostimulation.

So-called microcurrent therapy (MCT), which applies very weak currents (below the sensitivity threshold), cannot be therefore regarded as a strong stimulation; it works perhaps mainly as a placebo.

If two independent therapeutic circuits (at least two pairs of electrodes) of various but not widely different frequencies are applied simultaneously to the same body region, then this kind of electrostimulation is called interferential currents therapy. Classical interferential currents, also called Nemec's currents, are those oscillating around 4000 Hz; for example, one circuit applies 3900 Hz and another 4000 Hz or one circuit applies 4000 Hz and another 4100 Hz. In such a case, an interference of these currents results in creation of a low-frequency modulated current within the stimulated tissue. In practice this means that this classical interferential currents therapy has no any real advantage over the previously described modern, low-frequency TENS therapy. In general, nevertheless, it is good to combine two or more therapeutic circuits working on slightly different modern stimulation programs; their mutual overlapping will create an additional modulation and in this way could enhance the clinical efficacy of such an interferential TENS therapy.

"Softer" modulation of the TENS pulses is better tolerated by patients with acute conditions, such as renal colic or acute neuralgia, compared to more "hammering" sensations that are better suited for chronic problems; this is why modern TENS devices offer various combinations of modulated chains of pulses. To ensure user friendliness, instead of sophisticated technical descriptions, those more gentle stimulation programs should be simply labelled something such as "acute pain therapy," whereas those more irritating ones could be called something such as "chronic condition therapy." In the same way, "soft" electrostimulation programs suitable for the delicate skin of the face could be called "aesthetic mode" or "electric face lift," and those "stronger" ones, designed for the electrostimulation of muscles, could be called "slimming therapy" or "sports mode" (muscle electrostimulation is, however, a domain of the medical rehabilitation specialty). Nevertheless, the choice of more than four to six different therapeutic programs can only confuse the user and indicates purely commercial purposes of such designs aimed at the attraction of potential buyers. Specific electrostimulation programs for the hand, foot, upper back, lower back, etc. have no scientific foundations.

Typical TENS treatment comprises the use of one or more pairs of small, self-adhesive pad electrodes, square or, better, round, usually 2–5 cm in diameter, placed on the skin surface; usually local /OPAs/APs are used (Figure 5.8). The proper location of the electrodes is of utmost importance, because it determines the efficacy of treatment; all stimulated skin areas should display tenderness (practical, but subjective criterion) and a high degree of rectification/increased impedance phenomenon (objective scientific criterion) (see Sections 3.3 and 4.4.2). Intensity of the electrostimulation should be established according to the following rule: the stronger the better, but it must always be below the individual pain threshold, creating pleasant "tingling" sensations under the electrodes. To ensure equally strong sensations under both electrodes of the same channel, the electrodes should be of the same size and should be placed on similar kinds of skin (similar thickness and humidity), for example, symmetrically on both sides of the body. The usual treatment duration is 20–60 minutes per session, but if necessary it can be much longer; during labor pain, for example, electroanalgesia can be performed continuously for several hours. Depending on the seriousness of the treated condition, treatments can be repeated a few times per day, but should be done at

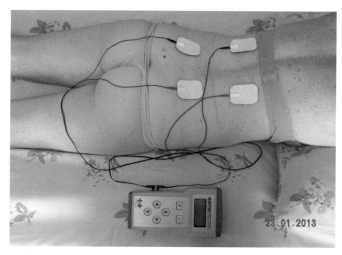

FIGURE 5.8 TENS treatment performed with a double-channel device (two pairs of electrodes).

least once per day when dealing with a chronic condition. There is no limitation regarding the number of sessions per therapeutic course; if necessary, TENS treatments can be done on a daily basis for a period of many years.

Typical indications for TENS therapy include arthritis, back pains of various origin, cervicogenic headache (including migraine), neuralgias (intercostal, trigeminal, and postherpetic), neuropathies (such as diabetic), reflex sympathetic dystrophy, Raynaud syndrome and other disturbances of peripheral blood circulation, allergic rhinitis, chronic sinusitis, bronchial asthma, spastic colon, renal colic, enuresis (children), atonic urinary bladder, atonic uterus (lack of hysterospasms), postoperative pains, menstrual pain, cancer pain, phantom pain, and toothache. TENS can be visibly effective as a supportive therapy for Parkinsonism; in such a case, the placement of electrodes can be similar to that for upper and lower back pain. Figure 5.9 illustrates the recommended placement of electrodes (two pairs) for common disorders and pain syndromes.

It is regrettable that certain manufacturers of TENS machines for the American market currently include in their instruction manuals the following statements: "TENS devices have no curative value," or "TENS does not cure any physiological problem, it only helps control the pain," or "TENS is a symptomatic treatment and as such suppresses the sensation of pain." These statements are required by the American Food and Drug Administration (FDA). However, these days it is common knowledge that TENS is of great therapeutic value even in nonpainful conditions, such as bronchospasm, rhinitis, Parkinsonism, and atonic urinary bladder/uterus; it can also accelerate the healing of wounds. Our own research studies (28, 66) in the field of diabetic neuropathy (widely regarded as "incurable") showed that a prolonged course of TENS leads to lasting clinical improvement, which includes not only pain relief but also a significant neurological improvement due to much better local blood microcirculation (see Chapter 3). Therefore, the previously mentioned statements have to be regarded as clearly false and should never be endorsed by a powerful certified body!

TENS treatment should not be applied to the chest in people with pacemakers, especially those received before 1999, nor to the lower back, abdomen, or legs in pregnancy. Electrodes

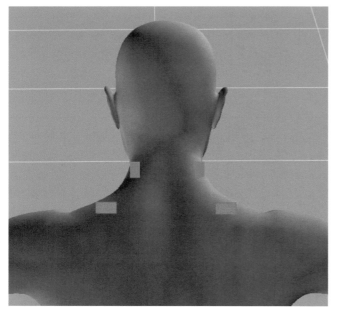
Upper back pain/Cervical syndrome

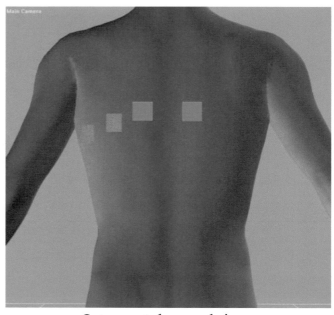
Intercostal neuralgia

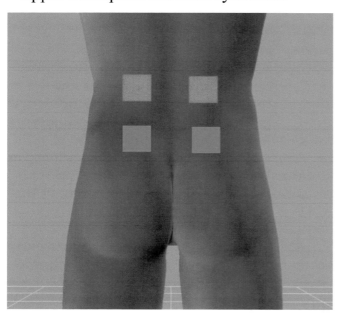
Lower back pain

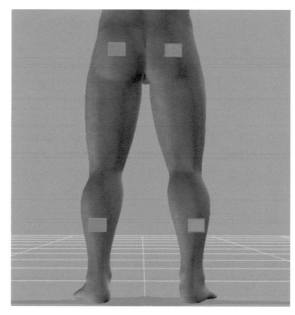
Sciatica (apply in addition to the lower-back treatment)

FIGURE 5.9 Recommended placement of TENS electrodes for common disorders and pain syndromes. Two channels, two pairs of pad electrodes: blue refers to channel 1 electrodes, red refers to channel 2 electrodes Please note: placements shown are only guidelines; sometimes better results can be achieved by moving electrodes according to individual needs.

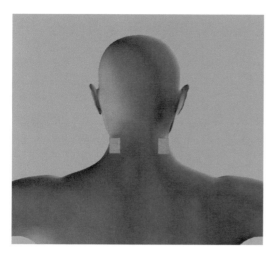
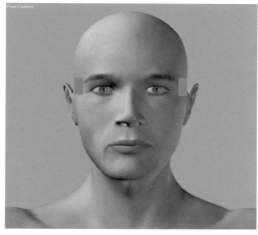

Headache (including migraine)

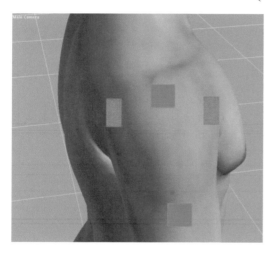
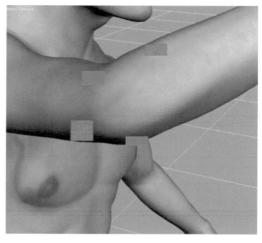

Shoulder pain

Elbow pain (tennis elbow)

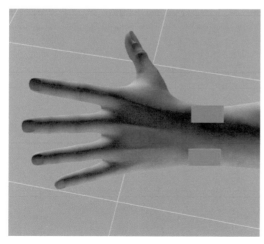
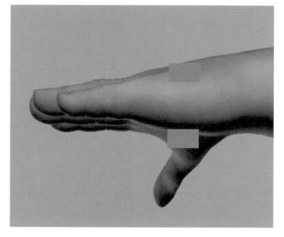

Painful wrist

Painful hand

FIGURE 5.9 cont'd

Raynaud syndrome

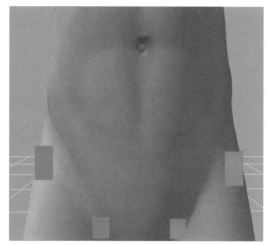
Painful hips

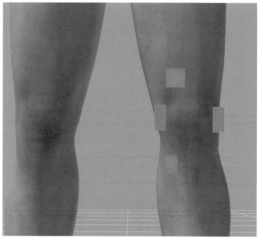
Knee pain

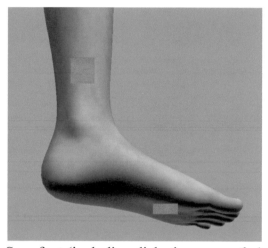
Sore foot (including diabetic neuropathy)

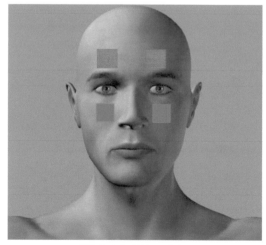
Rhinitis/Sinusitis

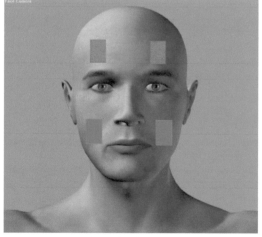
Electric face lift

FIGURE 5.9 cont'd

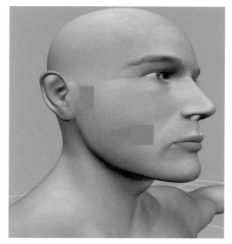
Facial palsy

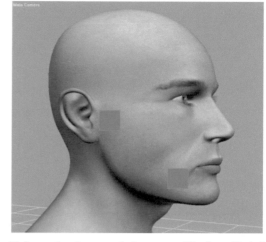
Trigeminal neuralgia: mandibular division

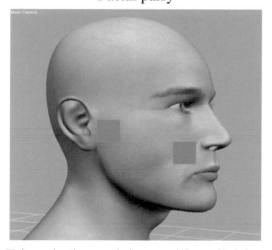
Trigeminal neuralgia: maxillary division

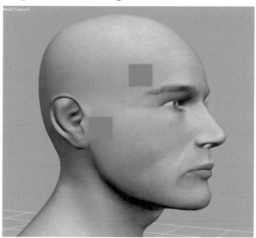
Trigeminal neuralgia: ophthalmic division

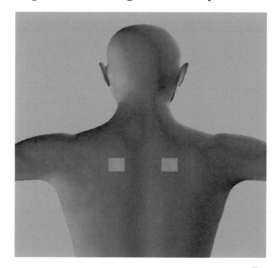
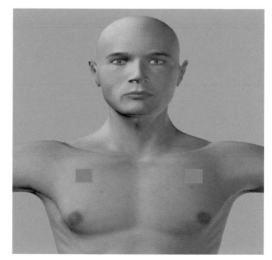
Bronchial asthma

FIGURE 5.9 cont'd

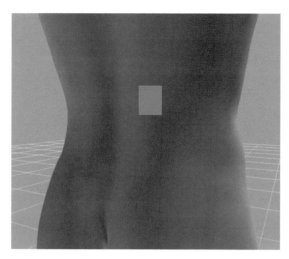
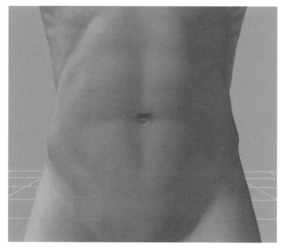

Renal colic

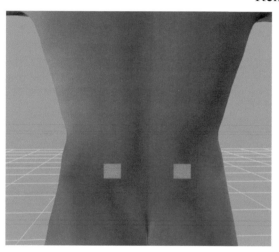
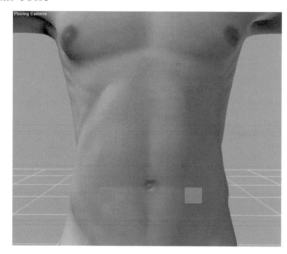

Spastic colon

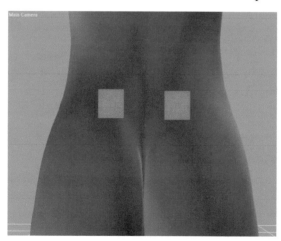
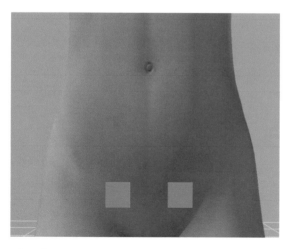

Menstrual pain

FIGURE 5.9 cont'd

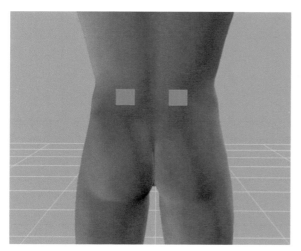
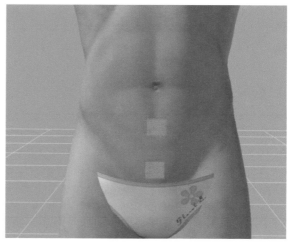

Nocturnal enuresis/Atonic urinary bladder

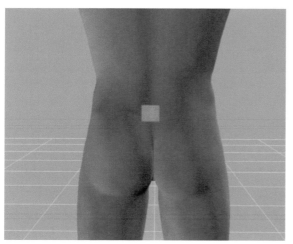
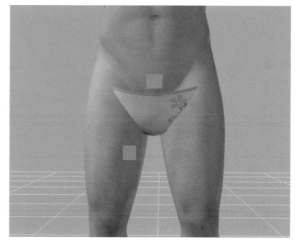

Impotence

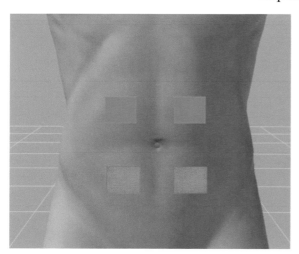
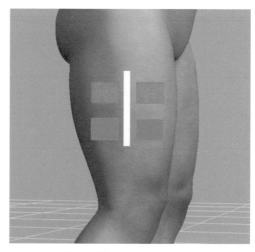

Slimming therapy   Cuts/Open wounds

FIGURE 5.9 cont'd

should not be placed on damaged skin because the stratum corneum is by far the most important electrical current barrier (see Section 2.1), any damage done to this layer will significantly increase local conductivity and therefore the density of electrical current passing through this area.

5.6.1.1 Electro-Sleep Therapy

So-called electro-sleep therapy is a version of the TENS treatment aimed at the patient's general relaxation. Electrodes are placed around the head; typically two electrodes are placed on the eyelids and another two behind the ears on the mastoideus regions (Figure 5.10). Instead of the eyelids, the electrodes are sometimes placed on both sides of the forehead or even on the ear lobes. In order to ensure a proper adhesion to the skin, especially on eyelids, thick layers of wet cotton wool are placed between the electrodes and the skin. "Eye electrodes" are usually in pairs with "mastoideus electrodes," so the currents cross the brain; this has nothing really to do with the electro-sleep mode of action, but it helps to adjust current intensities in such a way that vibrations are felt equally strongly on both eyelids. Preferential current parameters are: wave—bipolar impulse, impulse width—150–300 µs, frequency—between 3–20 Hz, pulse amplitude—should be individually adjusted to keep stimulation between the perception and the pain thresholds, no modulation. These parameters ensure monotonous stimulation, which contributes to the general sleepiness and relaxation in a manner similar to a monotonous train rumble. The usual duration of electro-sleep sessions is 45–90 minutes, but it is good to leave the patient asleep after the stimulation is switched off; the electro-sleep devices are, therefore, equipped with preprogrammed timers so that the patient does not need to be woken up. The course of treatment traditionally includes 15–20 sessions, but it can be repeated after one week.

Electro-sleep therapy can be helpful in stress, anxiety, and psychosis management as well as the management of drug addiction, motion sickness, and all the psychosomatic diseases, including insomnia. There are no direct contraindications for the use of this kind of electrotherapy.

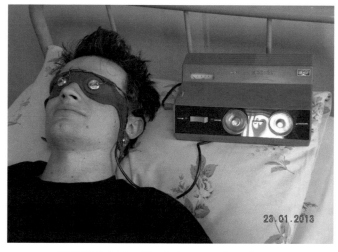

FIGURE 5.10 Classical electro-sleep therapy.

5.6.2 Direct Electrical Nerve Stimulation by Use of Implantable Pulse Generators

Electrostimulation of "quick" non-nociceptive nerve fibers, which can result in the blockage of painful signals carried by "slow" nociceptive nerve fibers (see Section 3.2), can be performed directly by means of pulse generators with leads implanted in the spinal cord. Relatively weak therapeutic currents will be still able to stimulate the thick non-nociceptive fibers, but not thin nociceptive ones.

In general, pulse waveforms generated by implantable devices are similar to those used for TENS purposes (see Figure 5.7). However, because direct nerve stimulation does not involve the epidermis, which is a powerful electrical current barrier, the pulse amplitude (peak voltage) is much lower: usually up to 10 V. The device can be implanted either in the upper abdomen area—typically below the rib cage—or in the subclavicular region. To ensure proper remote programming (telemetry), the implantable pulse generator should be placed not deeper than 4 cm beneath the surface of the skin. Proper location of leads in the spinal cord is essential for an optimal analgesic effect; detailed guidelines are provided in the manufacturers' instruction manuals. Yet the placement of leads in the wrong areas remains the main cause of a relatively high rate of unsuccessful clinical applications (Figure 5.11).

Direct electrical nerve stimulation by means of implantable pulse generators is mainly used in the management of severe pain syndromes of spinal origin. This provides the patient with convenient "on demand" access to electroanalgesia at any time and place. However, it requires a surgical intervention with a potential allergic or immune system response to the implanted materials. Also, the cost of the respective equipment is far higher than modern, portable TENS machines.

Electrostimulation of the lumbar spinal cord is contraindicated in pregnancy due to a potential risk of inducing premature delivery. Implantable analgesic pulse generators should not be used

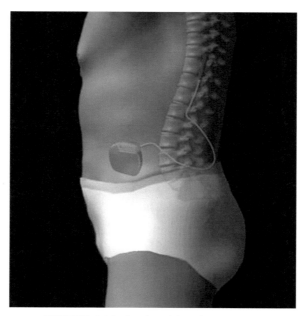

FIGURE 5.11 Implantable pulse generator.

in people with other implanted devices such as cardiac pacemakers. Spinal cord electrostimulation should be switched off when operating potentially dangerous equipment such as automobiles or power tools.

5.6.3 High-Frequency Electromagnetic Fields

Therapeutic procedures that use high-frequency electrical, magnetic, and electromagnetic fields are traditionally called diathermy: that is, deep heat. Depending on the wavelength and frequency of the electromagnetic vibrations, diathermy can be shortwave diathermy or microwave diathermy (longwave diathermy is no longer used in contemporary physical medicine). The respective parameters are regulated by the international convention as follows:

- So-called d'Arsonval's currents: frequency 300–500 kHz and wavelength 1000–600 m. (High frequency currents 1–5 MHz are used for the surgical diathermy)
- Electrical and magnetic fields utilized by a shortwave diathermy:
  - Frequency 13.56 MHz and wavelength 22.12 m
  - Frequency 27.12 MHz and wavelength 11.05 m
  - Frequency 40.68 MHz and wavelength 7.38 m
- Electromagnetic waves utilized by a microwave diathermy:
  - Frequency 433.92 MHz and wavelength 69.00 cm
  - Frequency 915.00 MHz and wavelength 32.80 cm
  - Frequency 2375.00 MHz and wavelength 12.62 cm
  - Frequency 2425.00 MHz and wavelength 12.4 cm

High-frequency electromagnetic vibrations can be used for therapeutic purposes in various ways:
- High-frequency current passing through the tissues, if the tissues are directly connected to the high-frequency electromagnetic vibrations generator
- High-frequency electrical field between two conducting plates of a condenser connected to the high-frequency electromagnetic vibrations generator
- High-frequency magnetic field, if the treated body region is placed inside or next to a coil connected to the high-frequency electromagnetic vibrations generator
- Electromagnetic field, that is, electromagnetic waves are created by the high-frequency electromagnetic vibrations generator

5.6.3.1 d'Arsonval's Currents

High-frequency currents are traditionally used as so-called d'Arsonval's currents: these are declining waves of high-frequency pulses (300–500 kHz), with short durations of such waves and 500 times longer intervals in between. The therapeutic procedure comprises the movement of special condenser electrodes along the treated body region, with flash-over discharges between the skin and the electrode surfaces. In the case of a very small distance between the electrode and the skin, the energy flow causes so-called dark discharges (effluvium). d'Arsonval's currents induce a limited amount of heat inside the tissues, which indicates a reflexive mode of action of this kind of therapy. These days, d'Arsonval's currents are widely used in cosmetology for skin rejuvenation purposes; otherwise, they can be used for the treatment of neuralgias, paraesthesias, pruritus, and frostbite.

5.6.3.2 Shortwave Diathermy

A shortwave diathermy overheats tissues by means of the high-frequency electrical or magnetic fields. Respectively, there are two methods of the shortwave diathermy:

- Condenser shortwave diathermy (high-frequency electrical field)
- Induction shortwave diathermy (high-frequency magnetic field)

Condenser shortwave diathermy (Figure 5.12) overheats the tissues by means of the high-frequency electrical field between two conducting plates of a condenser; these are two electrodes connected to the shortwave diathermy device. The electrodes can be stationary (usually 10 mm, 75 mm, or 128 mm in diameter), flexible (usually 120 × 180 mm or 180 × 240 mm), or of a special shape (for example, vaginal, axillary, or for boil treatment). Depending on the type of electrodes, they are always covered with various insulating materials for safety purposes: glass, rubber, plastics etc.

The amount of heat induced in tissues during the condenser shortwave diathermy treatment depends on the following factors:

- The size of the electrodes in relation to the treated body region
- The distance between the electrodes and the treated body region
- The type of the mutual electrode placement
- The type of dielectric between the electrodes and the body surface

Therefore, it is of utmost importance to strictly follow the instruction manuals of various shortwave diathermy devices in order to obtain optimal therapeutic results and to avoid any unwanted side effects.

Induction shortwave diathermy overheats the tissues by means of the high-frequency magnetic field. A high-frequency magnetic field can be applied to the treated body region either by placing this body region inside a coil or by exposing it to the influence of a dispersed magnetic field produced by the coil. The first method utilizes so-called cable electrodes, that is, flexible wires covered by a thick insulation mass; this kind of electrode can be wound

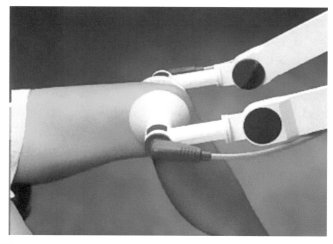

FIGURE 5.12 Condenser shortwave diathermy by means of stationary electrodes.

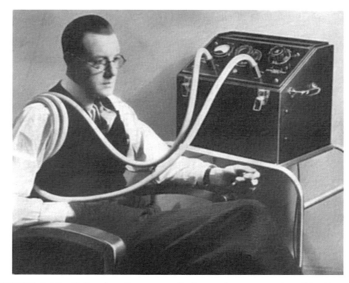

FIGURE 5.13 Induction shortwave diathermy by means of a cable electrode.

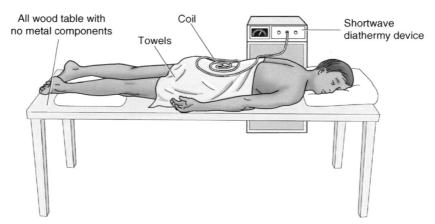

FIGURE 5.14 Induction shortwave diathermy.

around the treated body region such as the arm or the thigh (Figure 5.13). The second method uses so-called stationary induction electrodes, that is, coils placed inside an insulating cover; this kind of electrode can be fixed on the arm of a shortwave diathermy device (Figure 5.14). Induction shortwave diathermy overheats only superficially located tissues.

The amount of heat induced in tissues during the induction shortwave diathermy treatment depends on the following factors:

- Kind of tissue: more heat is produced in the high-conductivity tissues, for example, blood vessels or muscles, compared to the low-conductivity tissues like skin or fat tissue
- Coil current frequency: also, the number and shape of circumvolutions and position towards the treated body region
- Distance between the coil and the treated body region: a relatively longer distance allows deeper overheating

Contrary to various thermotherapeutic methods which apply so-called egzogenic heat from outside, shortwave diathermy produces an endogenic heat inside the tissues. A therapeutic dosage of shortwave diathermy can be estimated according to the following criteria:

- Individual patient's thermal sensations during the treatment:
  - Dosage I: athermic, slightly below the heat-perception threshold
  - Dosage II: oligothermic, very mild thermal sensations
  - Dosage III: thermic, nice thermal sensations
  - Dosage IV: hyperthermic, strong but not unpleasant thermal sensations
- Type and location of the pathological process:
  - The traditional rule is that in acute or subacute stages of a disease, mild dosages (I or II) are used, whereas in chronic pathologies, stronger dosages (III or IV) can be applied.
- Treatment duration:
  - Usually from five to 20 minutes depending on indications and dosage. Treatments can be done on daily or every other day. The traditional course of treatments consists of 15 to 20 sessions and can be repeated after one week.

The shortwave diathermy is a moderately effective method of physical medicine that can be widely used in all conditions for which the heat application can be beneficial, especially subacute and chronic inflammations. Some typical indications for shortwave diathermy treatment are described as follows:

- Chronic arthritis: induction or condenser electrodes (distance from the treated body region, respectively 2–4 cm for the active electrode and 2–4 cm for the passive electrode), dosage II–III, treatment duration 5–10 minutes per joint.
- Back pains: induction or condenser electrodes (distance from the treated body region, respectively 2–4 cm for the active electrode and 2–4 cm for the passive electrode), dosage II–III, treatment duration 15–20 minutes
- Neuralgia/Neuropathy: condenser electrodes (distance from the treated body region, respectively 2–4 cm for the active and 2–4 cm for the passive electrode), dosage II–IV, treatment duration 15–20 minutes
- Tendonitis: condenser electrodes (distance from the treated body region, respectively 2 cm for the active and 4 cm for the passive electrode), dosage II–III, treatment duration 5–10 minutes
- Chronic sinusitis: condenser electrodes (distance from the treated body region, respectively 2–4 cm for the active and 2–4 cm for the passive electrode), dosage II, treatment duration 10–15 minutes
- Chronic otitis: condenser electrodes (distance from the treated body region, respectively 1–3 cm for the active and 4–6 cm for the passive electrode), dosage II, treatment duration 10–15 minutes
- Chronic tonsillitis: condenser electrodes (distance from the treated body region, respectively 3 cm for the active and 3 cm for the passive electrode), dosage II, treatment duration 10–15 minutes
- Chronic laryngitis: condenser electrodes (distance from the treated body region, respectively 1–3 cm for the active and 1–3 cm for the passive electrode), dosage II–III, treatment duration 10–15 minutes
- Bronchial asthma: condenser electrodes (distance from the treated body region, respectively 6 cm for the active and 6 cm for the passive electrode), dosage II–III, treatment duration 10–15 minutes

- Chronic obstructive pulmonary disease (COPD): condenser electrodes (distance from the treated body region, respectively 4 cm for the active and 4–10 cm for the passive electrode), dosage II–IV, treatment duration 15–20 minutes
- Chronic colitis: condenser electrodes (distance from the treated body region, respectively 2–3 cm for the active and 3 cm for the passive electrode), dosage II–III, treatment duration 10–15 minutes
- Chronic pyelonephritis: condenser electrodes (distance from the treated body region, respectively 4–6 cm for the active and 6 cm for the passive electrode), dosage II–III, treatment duration 10–15 minutes
- Chronic prostatitis: condenser electrodes (distance from the treated body region, respectively 2–4 cm for the active and 3–5 cm for the passive electrode), dosage II–III, treatment duration 10–15 minutes
- Chronic adnexitis: condenser electrodes (distance from the treated body region, respectively 2–4 cm for the active and 4–6 cm for the passive electrode), dosage II–III, treatment duration 15–20 minutes
- Ovarian endocrynological dysfunction: condenser electrodes (distance from the treated body region, respectively 2–4 cm for the active and 6 cm for the passive electrode), dosage II–III, treatment duration 10–15 minutes
- Mastitis: induction or condenser electrodes (distance from the treated body region, respectively 3–5 cm for the active and 4–6 cm for the passive electrode), dosage III, treatment duration 5–10 minutes
- Frostbite: condenser electrodes (distance from the treated body region, respectively 2–3 cm for the active and 2–3 cm for the passive electrode), dosage II–III, treatment duration 10–15 minutes

During the shortwave diathermy procedure, all nearby metal objects should be removed; the treatment must be done on wooden, not metal, couches. The cables connecting the therapeutic device with electrodes must not touch the skin due to the risk of burns; in such a case, a piece of felt should be used as an insulation. If two surfaces of the heated body region touch each other, for example, the thighs, they must also be separated by means of a piece of felt in order to prevent potential burns. The patients should be undressed, because the wet cloth can be overheated (also wet bandages); they must not move nor touch the therapeutic device.

Shortwave diathermy should not be used in case of malignant pathology, tuberculosis, septic infections (abscess), deep vein thrombosis (DVT), or pregnancy. People with various metal implants, especially pacemakers, are not allowed to be in the vicinity of high-frequency electromagnetic wave generators. For the same reason, shortwave diathermy cannot be combined with acupuncture in one session.

5.6.3.3 Pulsed High-Frequency Magnetic Field Therapy

This is a newer version of the induction shortwave diathermy, characterized by chains of high-pick-power impulses. The intervals between impulses are long enough to have disperse heat, yet this kind of therapy is of proven therapeutic efficacy. This clearly indicates a reflexive mode of action of high-frequency electromagnetic field therapies. The usual impulse

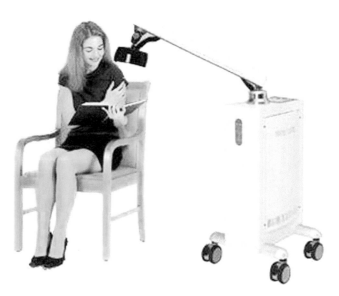

FIGURE 5.15 Pulsed high-frequency magnetic field therapy.

duration is a μs range (for example, 60 or 100 μs) and the usual frequency range is between several and a few hundred Hz. The impulse pick-power (so-called penetration) usual range is between 300 and 1000 W.

The pulsed high frequency magnetic field therapy (Figure 5.15) is especially useful in the management of various inflammations, edemas, and pains. Therefore, indications for this kind of therapy include arthritis, soft tissue injuries, slowly healing wounds and trophic ulcerations, chronic sinusitis, chronic adnexitis, and certain skin diseases.

5.6.3.4 Microwave Diathermy

Microwave diathermy (Figure 5.16) overheats tissues by means of the microwave-frequency electromagnetic field. Microwaves are electromagnetic waves of a wavelength from 0.1 to 100 cm. They can be generated by a special generating lamp called a magnetron. Microwaves' characteristics are different from those of radiowaves and similar to infrared and visible radiation. Therefore, they undergo reflection, dispersion, refraction, and diffraction on various complicated tissue structures. About 50% of the microwave beam applied to the skin is reflected; the rest is absorbed by tissues at depths of only 6–8 cm. Microwaves cause an oscillation of ions in electrolytes and molecules in polarized dielectrics and in this way create heat. The most overheated are tissues containing a lot of water, such as blood and muscles; fat tissue containing little water is only slightly affected.

Microwaves are applied to the body by means of radiators that are connected to the magnetron. Radiators can be of different sizes depending on the kind and location of pathology. They are usually placed 5–10 cm from the skin surface. The microwave dosages are estimated in a similar way to shortwave diathermy (I, II, III, and IV). Another classification, based on watts, includes mild dosages (up to 20 W) and strong dosages (up to 150 W): the most often-used dosages are those between 20 and 100 W. Treatment duration, depending on indications, is 5–15 minutes. The usual course of treatments includes 10–15 sessions.

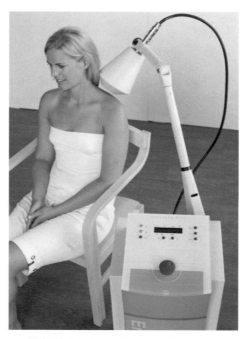

FIGURE 5.16 Microwave diathermy.

Indications for microwave diathermy are generally similar to those for shortwave diathermy. A superficial penetration of microwaves should not constitute any limitation, because a reflexive mechanism seems to be a leading mode of action for both methods. Therefore this kind of therapy can still be widely used for, among other conditions, arthritis, back pains, neuralgias, and (using a special radiator) chronic adnexitis.

Contraindications are the same as in the case of shortwave diathermy. It is also important to remember that microwaves can be harmful; eye lenses and the reproductive tissues of testicles and ovaries are especially at risk. Partial protection can be provided by screening the room in which the microwave diathermy is performed.

5.7 MAGNETOTHERAPY

Magnetic fields, just like electromagnetic ones, are omnipresent in nature. Magnetic phenomena are strictly connected with electrical phenomena and vice versa. Thus, alternating electric fields generate alternating magnetic fields, and alternating magnetic fields generate alternating electric fields with eddy currents induced in conductors. Magnetic field intensity can be estimated in magnetic induction units: $1T$ (Tesla) $= 1V \times s/m^2$.

Over the centuries, various attempts have been undertaken to use magnetic fields for therapeutic purposes. However, despite a lot of interest, there is still not much known about the biological influence of magnetic fields. In contrast to other energy forms that are absorbed at certain depths, magnetic fields can penetrate throughout all structures of the body (even through the clothes), but the applied energy is very low. This indicates a stimulating, most likely reflexive mode of action of magnetotherapy.

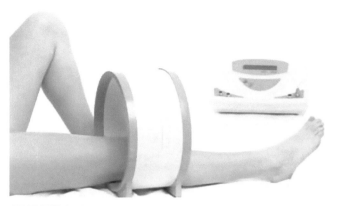

FIGURE 5.17 ELFMF therapy by means of a solenoid applicator.

Magnetotherapy belongs to the "soft" methods of physical medicine, with its therapeutic effectiveness being generally lower than other physical therapies. Still, it can be successfully used in mild cases, and in severe problems it can be an important supportive treatment. Initially, stationary magnets of various magnitudes were applied to diseased body parts. More recently, small magnets have also been attached to APs, especially auricular OPAs. However, as is the case with other reflexive methods, intermittent/modulated therapy displays better stimulating characteristics. Therefore, contemporary physical medicine uses various generators emitting pulsed magnetic fields. Of special interest is the so-called extremely low-frequency magnetic fields (ELFMF) therapy with a frequency ranging from 1 to 60 Hz and magnetic induction from 0.1 to 20 mT. Another direction of the therapeutic utilization of magnetic fields is the so-called microTesla magnetic field therapy, which proposes lower induction (up to 100 µT, approximate to the terrestrial magnetic field induction) but higher frequency, up to 3000 Hz.

By general analogy to electrotherapy, in order to insure a maximal therapeutic effect of ELFMF therapy, it is good to use the most "stimulating" parameters of the applied magnetic fields: shape—bipolar square, frequency—3–20 Hz, induction—at least 10 mT, modulation—all parameters. It is also good to apply an intermittent stimulation rather than continuous; recommended intervals are one to two seconds. There are two types of applicators: solenoids (Figure 5.17), usually 20–60 cm in diameter, or flat applicators (Figure 5.18), to be placed next to the treated body parts. Flat applicators can be in the form of mattresses or pillows. The usual treatment duration is 20–60 minutes; treatments can be done on a daily basis or at least every other day. The course of treatments consists usually of 12–20 sessions, but it can be repeated as often as needed. As indicated earlier, patients do not need to take off their clothes.

There are wide indications for the use of ELFMF therapy, including arthritis, back pains, neuralgias, soft tissue injuries, disturbances of the peripheral blood circulation, slowly healing bone fractures, wounds, and trophic ulcerations.

There are no direct contraindications for ELFMF therapy, but it is suggested that this kind of treatment should not be used in malignancy, pregnancy, active pulmonary TB, or for patients with electronic implants. Metallic prostheses are not contraindications for magnetotherapy.

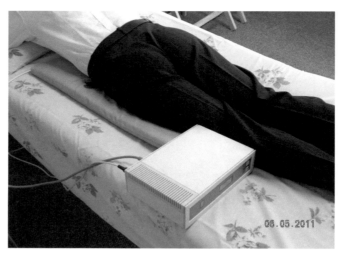

FIGURE 5.18 ELFMF therapy by means of a flat applicator in the form of a mattress.

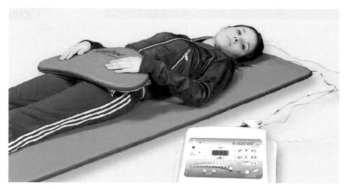

FIGURE 5.19 MicroTesla magnetic fields therapy.

MicroTesla magnetic fields therapy is a relatively new version of magnetotherapy, and more research is needed to estimate its clinical efficiency and scope of utilization. The idea is based on the controversial ion cyclotronic resonance theory, which suggests that various tissues (in our case, nervous tissue) respond especially well to the magnetic field stimulation of specific induction and frequency parameters. Technically, in the case of the higher basic frequency of the microTesla magnetic fields, the chains of impulses are usually divided as packs with a frequency of several Hz. Applicators are in the form of mattresses and pillows, to be placed under the treated body parts (Figure 5.19). The usual treatment duration is 20–60 minutes; treatments can be done on a daily basis or at least every other day. The course of treatments usually consists of 12–20 sessions, but it can be repeated as often as needed. Patients do not need to take off their clothes.

Indications for the microTesla magnetic fields therapy are generally similar to the ELFMF therapy; however, there are promising signs that this type of treatment also can be helpful in Parkinson's as well as Alzheimer's diseases, multiple sclerosis, brain injuries, and migraines. There are no known contraindications for the use of microTesla magnetic fields therapy.

5.8 REFLEXIVE MECHANICAL STIMULATION

Auto-massage of painful body parts comes as a natural reflex after any injury. Therefore, therapeutic massage is perhaps the oldest medical procedure and is known in the oldest civilizations. However, the ancient Far Eastern doctors developed a specific kind of point massage (Chinese Tien-An or Japanese Shiatsu) aimed at the stimulation of trigger/pressure points (Ashi points): these days, this reflexive massage is often called pressopuncture or acupressure, which are not very appropriate terms from a linguistic point of view. To make the stimulation even more effective, Far Eastern doctors used needles (acupuncture), heat (moxibution; see Section 5.1.1), and, later, cupping (see Section 5.8.5). Some contemporary therapists also use small skin incisions/stitches ("prolonged acupuncture"), certain creams/plasters, and even subcutaneous injections of various substances ("Prolotherapy," etc.); clear water or certain herbal remedies are very strong stimuli of the skin nervous receptors, but injections of, for example, vitamins or other supplements do not make sense. Bee stings are perhaps the most powerful stimuli, but their therapeutic use is very limited due to the risk of anaphylactic shock; such a procedure can also be very painful. Electrical stimulation of acupuncture needles significantly increases the efficacy of such a combined treatment, as does thermotherapeutic stimulation with moxa or, better, cryotherapy. Injections of local anesthetics, such as 1–2% Lignocaine, act as peripheral nerve blocks and should not be classified as a typical physical medicine.

In the following sections, the most important systems of reflexive mechanical stimulation will be discussed.

5.8.1 Acupuncture ("Dry Needling")

Acupuncture is arguably the most effective and versatile method of physical medicine, and there are hundreds of research articles, published in various peer-review medical journals, that confirm its therapeutic efficacy in a wide variety of pathological conditions. There are even certain medical journals exclusively dedicated to acupuncture therapy; among the most popular are *Deutsche Zeitschrift fur Akupunktur, Medical Acupuncture, the Journal of Clinical Acupuncture, and the Journal of Traditional Chinese Medicine*. Yet acupuncture is still not recognized as an official therapy by certain academic societies and generally is not included in "Western" manuals of physical medicine despite the fact that it fulfills all the criteria to be classified as such. This is due to the lack of a clear physiological explanation of acupuncture's mode of action that is in accordance with a contemporary state of medical knowledge.

It seems that the extraordinary clinical effectiveness of acupuncture is based on two factors: therapeutic inclusion of distal OPAs (contrary to Western physical therapies, which usually concentrate on the affected body areas) and the use of very strong stimuli, that is, the needle punctures. The convergence modulation theory (see Section 3.2) provides an acceptable explanation of the neurophysiological mechanisms that are the basis for the acupuncture therapy, and the use of OED allows a scientific selection of the most suitable APs (see Sections 4.4.2 and 5.1).

5.8.1.1 History

The 6000-year-old traditional Chinese medicine is based on the unique dualistic cosmic theory of the Yin and the Yang. The Yang, the male principle, is active and light (the sunrise, heat, or a clear day); the Yin, the female principle, is passive and dark (the sunset, cold, or a cloudy day). The human body, like matter in general, is made up of five elements: wood, fire, earth, metal, and water. Health, character, and even the success of any ventures are determined by the preponderance at any time of either the Yin or the Yang. In the body, their proportions can be controlled; this was the great aim of ancient Chinese medicine, to be achieved with meditation, exercises, physical therapies, and herbal remedies. It must be emphasized that traditional Chinese medicine did not concentrate on only one therapeutic method such as acupuncture or herbal remedies alone but adopted a holistic approach with various therapies used according to actual needs.

Because autopsies in ancient China were forbidden, traditional Chinese anatomy is based on the cosmic system, which postulates the presence of such hypothetical structures as the 12 channels (main meridians) of the "vital energy" Qi. In general, the meridians roughly follow the typical pain radiation of the real internal organs that they are passing through. Traditional anatomical knowledge is, however, a corroboration of the system; when there is a contradiction, the system is assumed to be right. The body contains five Yin solid organs (heart, lungs, liver, spleen, and kidneys), which store up but do not eliminate; and five Yang hollow viscera (stomach, small intestine, large intestine, gallbladder, and urinary bladder), which eliminate but do not store up.

According to the physiology of traditional Chinese medicine, the blood vessels are supposed to contain blood and air in proportions varying with those of the Ying and the Yang. These two cosmic principles circulate in the 12 channels and control the blood vessels and hence the pulse; the assessment of the pulse is therefore the most important part of the traditional examination. Traditional examination nevertheless also includes a strict evaluation of the patient's tongue, an assessment of smell, and historically a detailed palpation of the skin; in this way, perhaps, the phenomenon of an increased tenderness of the APs related to diseased organs was discovered.

Traditional Chinese pathology is also dependent on the theory of the Yin and the Yang; this led to an elaborate classification of diseases in which most of the types listed have no scientific foundation. Health depends on the harmonious balance of the Yin and the Yang. If the flow of one of these principles is obstructed, disharmony and disease result. Acupuncture needles, inserted into the right skin spots, are believed to affect the distribution of the Yin and the Yang in the meridians and thereby restore the required "energetic" balance. It comes as a surprise, however, that such an unmeasurable enigmatic energy as the "vital energy" Qi could be controlled by means of such primitive tools as pieces of wood/bamboo or stone/metal needles.

There are several complicated traditional rules, described by 2500 BC, concerning the selection of appropriate APs for the treatment of various conditions. If all these rules are to be applied, then the total number of needles used for an average acupuncture treatment would be counted in the dozens. Yet different authors recommend stimulation of very different APs for the same clinical problem.

In the second half of the 20th century, despite much clinical evidence in favor of acupuncture, chemically orientated Western academic medicine promptly rejected this "nonscientific method" because there were no obvious chemical foundations. In contrast, "alternative" practitioners who were generally not highly qualified but were very enthusiastic directly recognized the hypothetic "vital energy" as electricity, laser, or cosmic radiation. Various strange measuring devices have been built and sometimes sold in large numbers (see Sections 2.5.1 and 2.7); these devices are supposed to measure the bio-energy circulating in the human body, but how is that possible if we do not even know the units of bio-energy? This situation created a polarization of attitudes: Any rational scientific research in the field cannot gain any real interest from either academic societies ("another crazy healer with another crazy idea") or from traditional or "alternative" circles (they prefer their simple "wishful thinking").

Figures 5.21–5.34 illustrate topography of the 12 main meridians and classical APs; the included description of the APs' therapeutic indications is strictly traditional. In order to address individual differences, when it comes to the localization of APs, traditional Chinese medicine introduced a proportional unit of measurement called cun. This is a distance between the second and the third joint of the third finger; it is also the width of the thumb (Figure 5.20). These days, meridians should be understood as artificial lines connecting APs that are roughly related to the same organs/body parts (see Section 2.7); in contemporary practice, meridians are useful for the localization of particular APs.

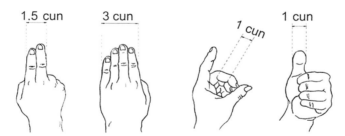

FIGURE 5.20　Cun: a proportional unit of measurement used in traditional Chinese medicine.

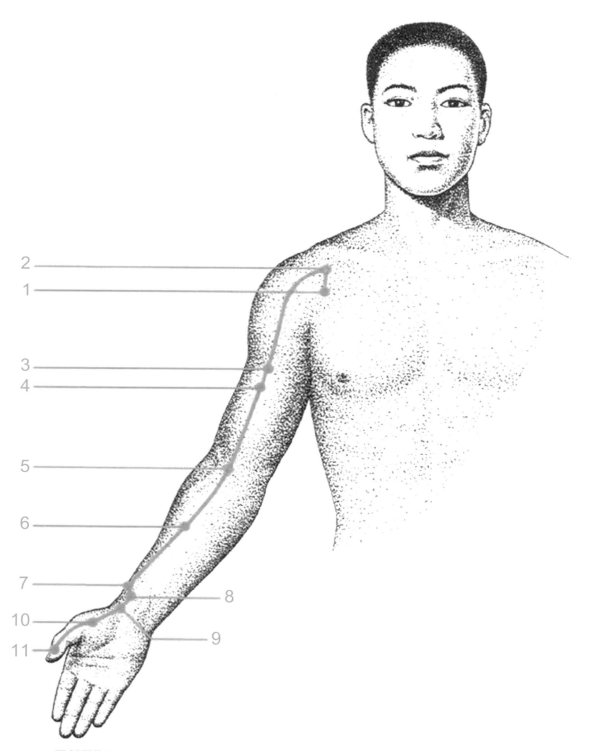

FIGURE 5.21A   Typical topography of the lung meridian (L) according to contemporary sources.

FIGURE 5.21B Topography of the lung meridian according to the Chinese atlas Shih Ssu Ching Fa Hui from the year 1341.

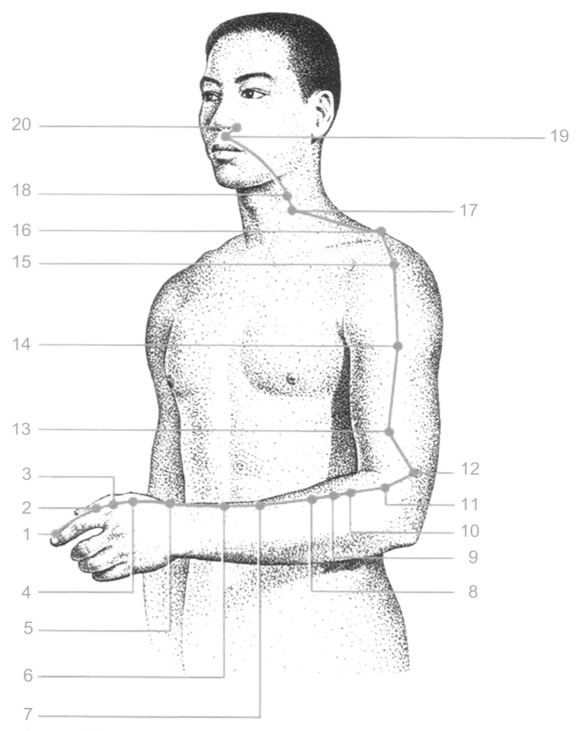

FIGURE 5.22A Typical topography of the large intestine meridian (LI) according to contemporary sources.

5.8 REFLEXIVE MECHANICAL STIMULATION

FIGURE 5.22B Topography of the large intestine meridian according to the Chinese atlas Shih Ssu Ching Fa Hui from the year 1341.

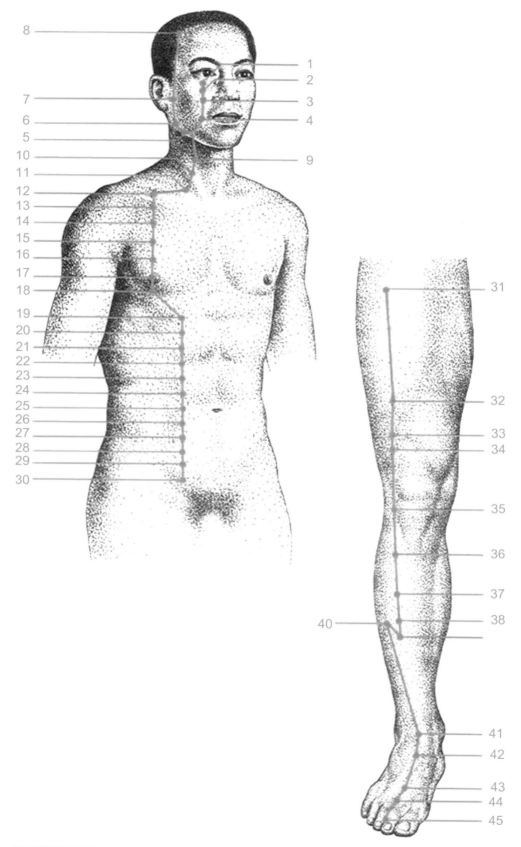

FIGURE 5.23A Typical topography of the stomach meridian (S) according to contemporary sources.

5.8 REFLEXIVE MECHANICAL STIMULATION

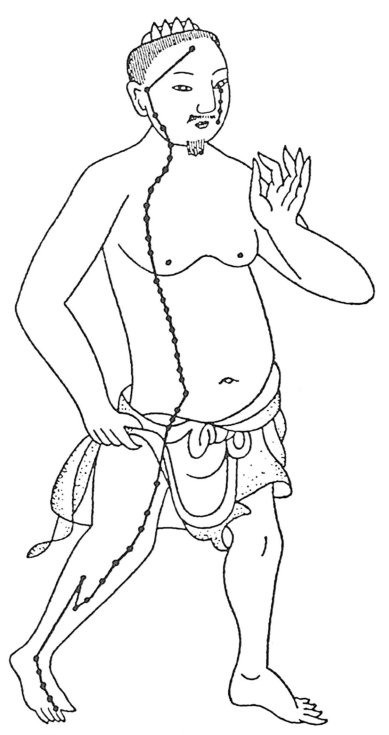

FIGURE 5.23B Topography of the stomach meridian according to the Chinese atlas Shih Ssu Ching Fa Hui from the year 1341.

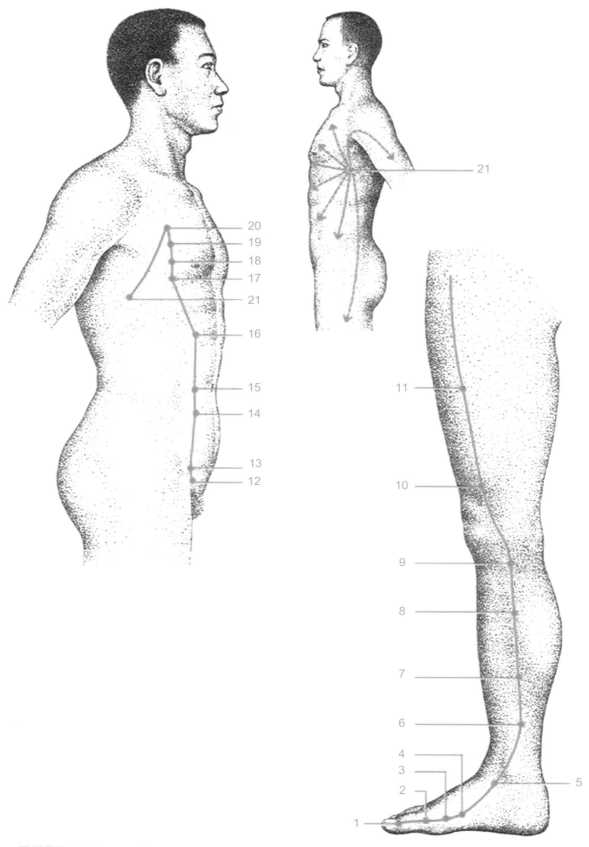

FIGURE 5.24A   Typical topography of the spleen and pancreas meridian (SP) according to contemporary sources.

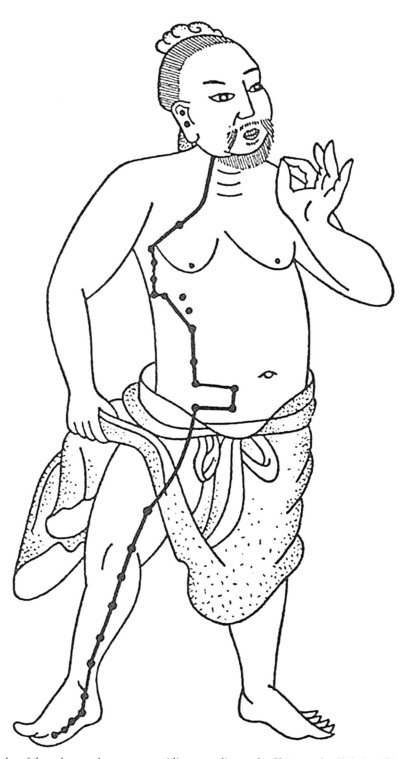

FIGURE 5.24B Topography of the spleen and pancreas meridian according to the Chinese atlas Shih Ssu Ching Fa Hui from the year 1341.

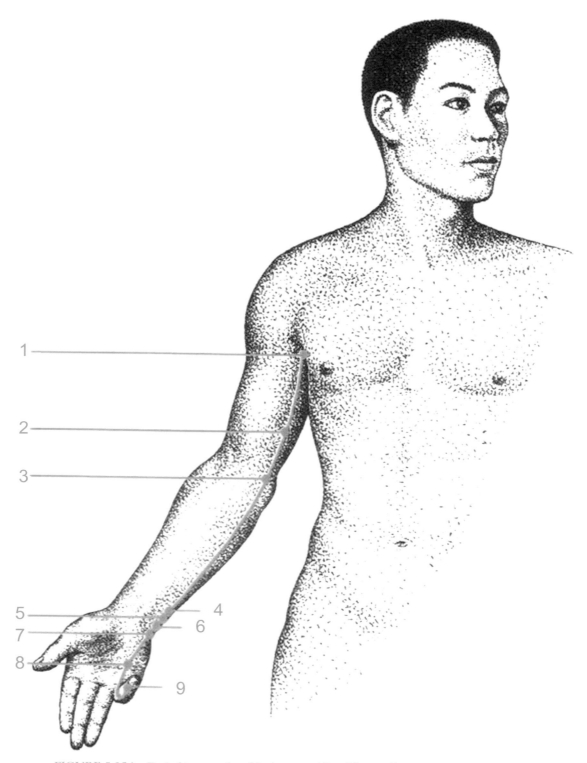

FIGURE 5.25A Typical topography of the heart meridian (H) according to contemporary sources.

5.8 REFLEXIVE MECHANICAL STIMULATION

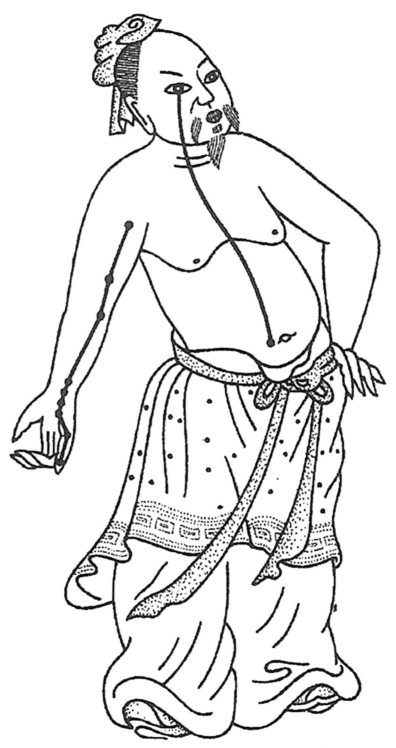

FIGURE 5.25B Topography of the heart meridian according to the Chinese atlas Shih Ssu Ching Fa Hui from the year 1341.

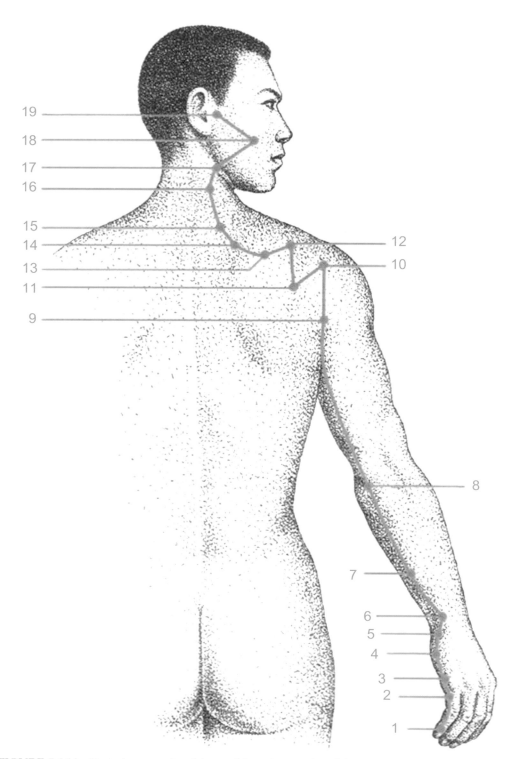

FIGURE 5.26A Typical topography of the small intestine meridian (SI) according to contemporary sources.

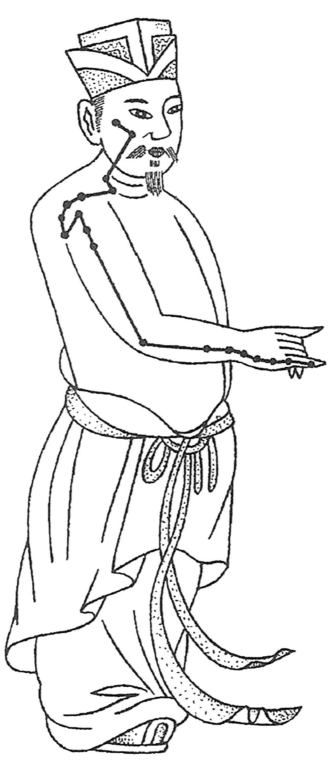

FIGURE 5.26B Topography of the small intestine meridian according to the Chinese atlas Shih Ssu Ching Fa Hui from the year 1341.

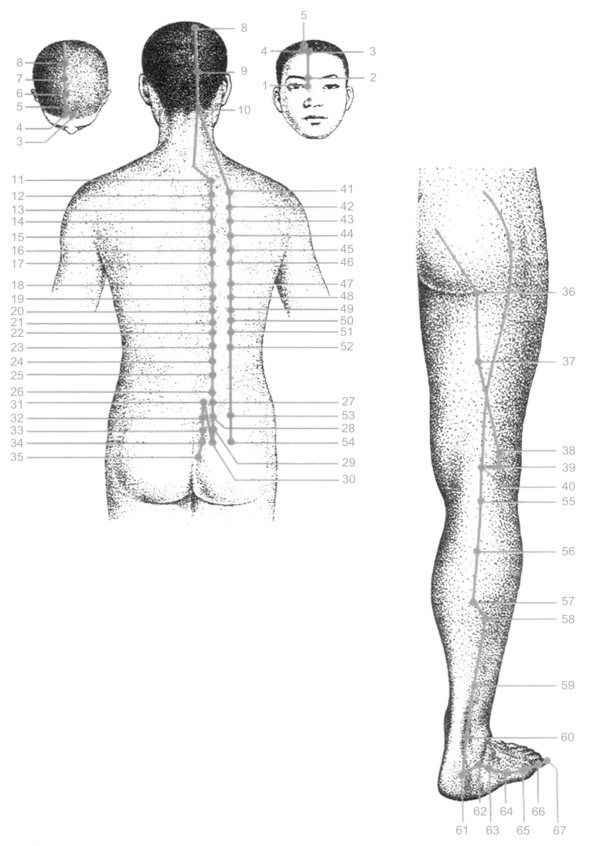

FIGURE 5.27A Typical topography of the urinary bladder meridian (B) according to contemporary sources.

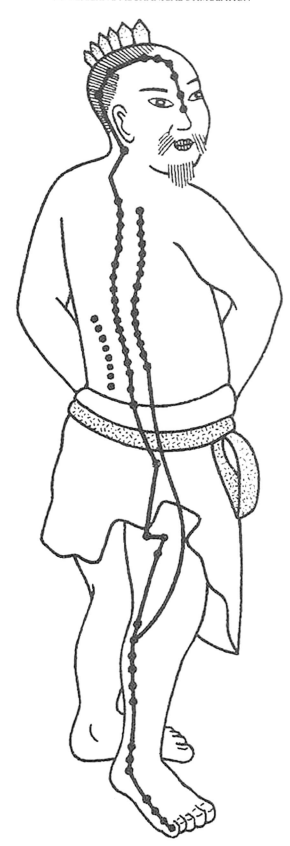

FIGURE 5.27B Topography of the urinary bladder meridian according to the Chinese atlas Shih Ssu Ching Fa Hui from the year 1341.

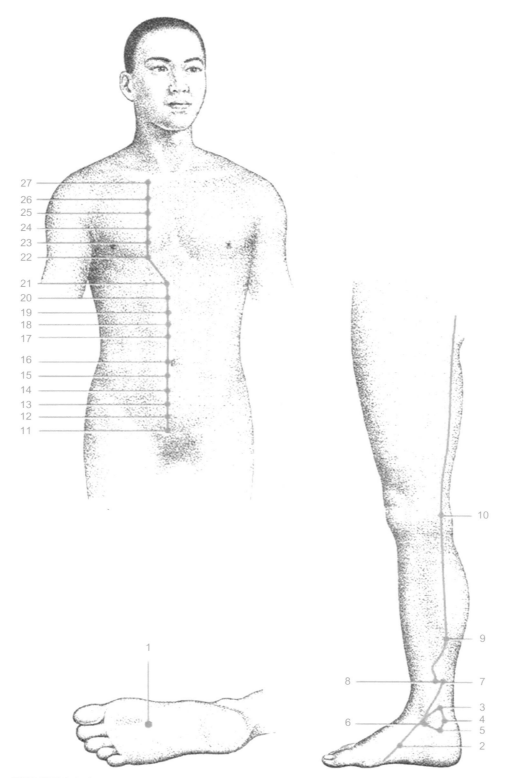

FIGURE 5.28A Typical topography of the kidney meridian (K) according to contemporary sources.

FIGURE 5.28B Topography of the kidney meridian according to the Chinese atlas Shih Ssu Ching Fa Hui from the year 1341.

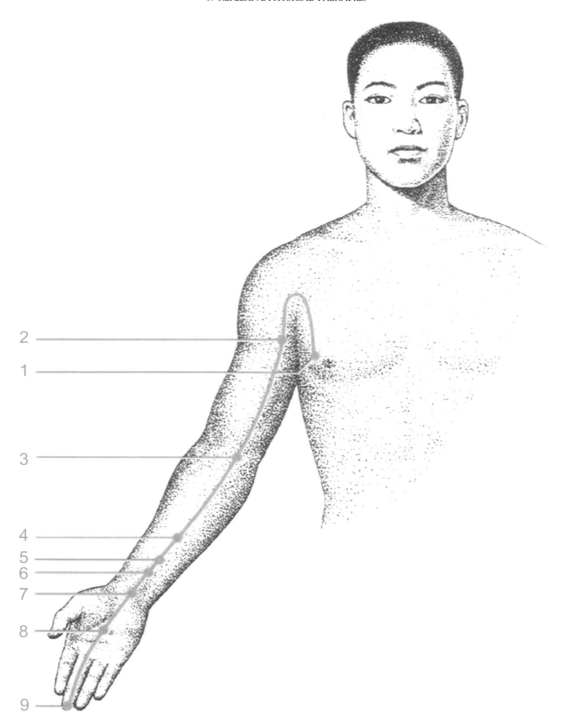

FIGURE 5.29A Typical topography of the pericardium meridian (P) according to contemporary sources.

FIGURE 5.29B Topography of the pericardium meridian according to the Chinese atlas Shih Ssu Ching Fa Hui from the year 1341.

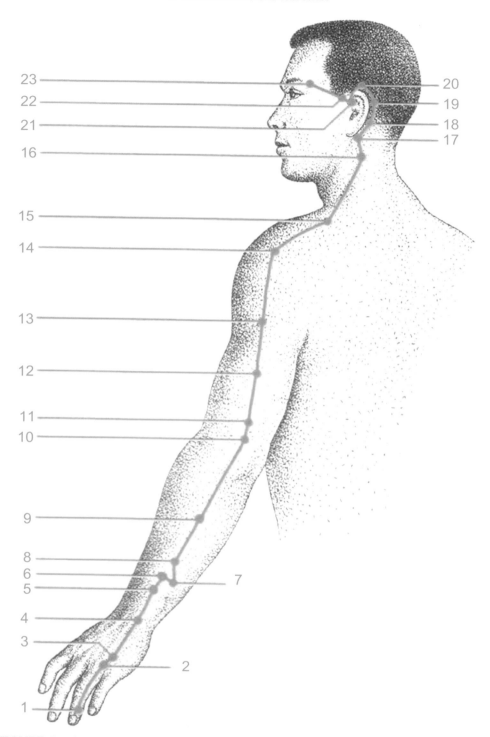

FIGURE 5.30A Typical topography of triple warmer meridian (T) according to contemporary sources.

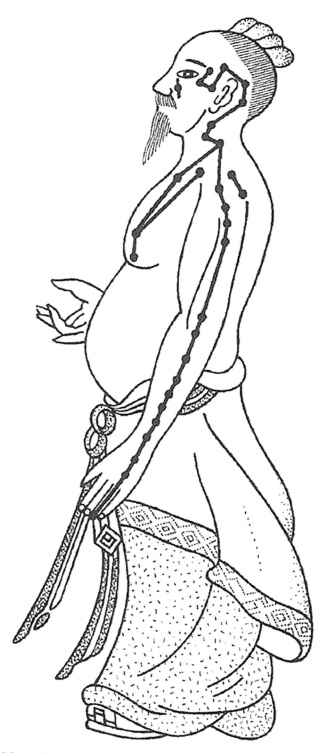

FIGURE 5.30B Topography of the triple warmer meridian according to the Chinese atlas Shih Ssu Ching Fa Hui from the year 1341.

132　　5. REFLEXIVE PHYSICAL THERAPIES

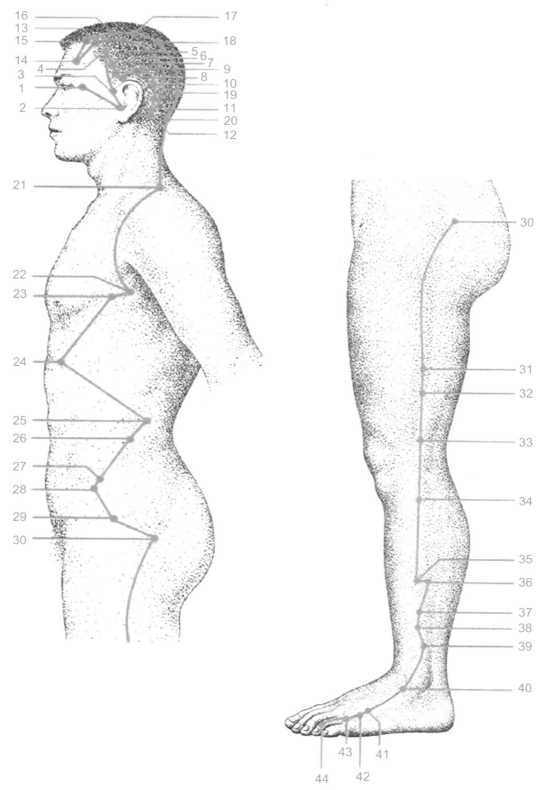

FIGURE 5.31A　Typical topography of the gallbladder meridian (G) according to contemporary sources.

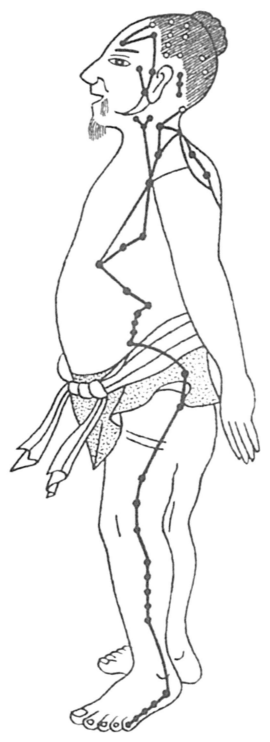

FIGURE 5.31B Topography of the gallbladder meridian according to the Chinese atlas Shih Ssu Ching Fa Hui from the year 1341.

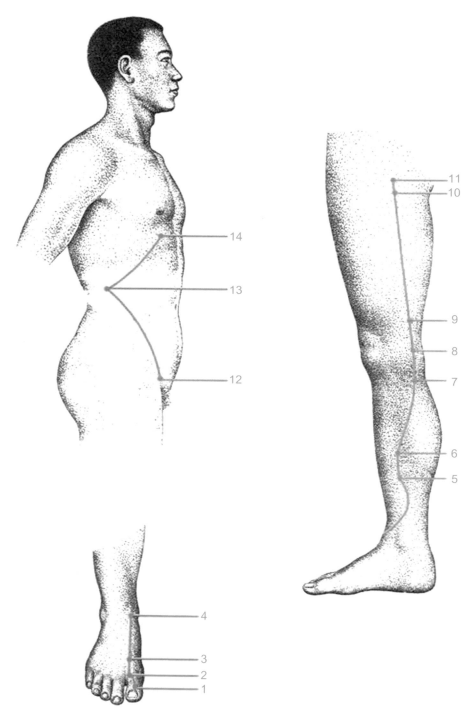

FIGURE 5.32A Typical topography of the liver meridian (Liv) according to contemporary sources.

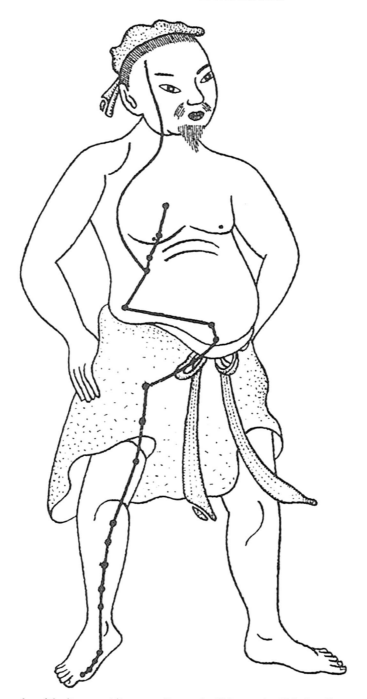

FIGURE 5.32B Topography of the liver meridian according to the Chinese atlas Shih Ssu Ching Fa Hui from the year 1341.

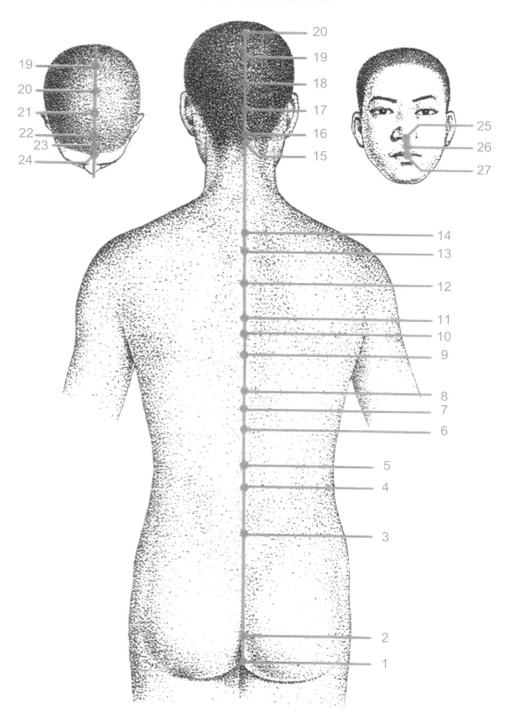

FIGURE 5.33A Typical topography of the governing vessel (GV) according to contemporary sources.

5.8 REFLEXIVE MECHANICAL STIMULATION

FIGURE 5.33B Topography of the governing vessel according to the Chinese atlas Shih Ssu Ching Fa Hui from the year 1341.

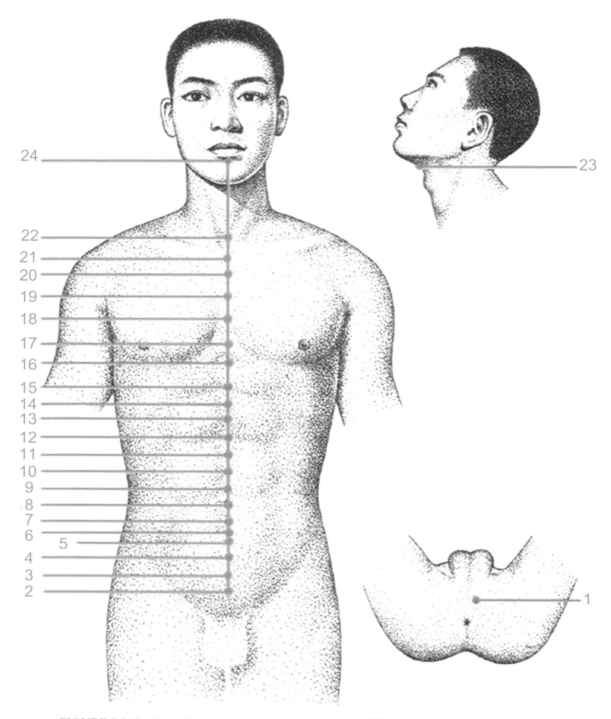

FIGURE 5.34A Typical topography of the conception vessel (CV) according to contemporary sources.

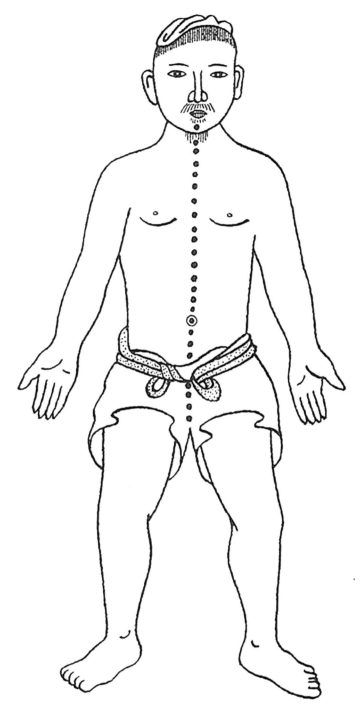

FIGURE 5.34B Topography of the conception vessel according to the Chinese atlas Shih Ssu Ching Fa Hui from the year 1341.

CLASSICAL APs PERTAINING TO THE LUNG MERIDIAN (L)

Point: Zhonfu (L1)
Location: 1 cun below Pt. Yunmen, in first intercostal space.
Indications: Cough, asthma, pain in the chest, fullness of the lung, pain of the shoulder and the back.

Point: Yunmen (L2)
Location: 6 cun lateral to the midline of the chest, at the level of the lower border of the clavicula, when the upper extremity is flexed in position it locates in the depression in the infraclavicular fossa.
Indications: Cough, asthma, fullness of the chest, pain in the chest, etc.

Point: Tianfu (L3)
Location: 6 cun above Pt. Chize on the radial side of m. biceps brachi.
Indications: Asthma, epistaxis, pain in the medial aspect of the upper arm.

Point: Xiabai (L4)
Location: 5 cun above Pt. Chize on the radial side of m. biceps brachi.
Indications: Cough, dyspnea, fullness of the chest, pain in the medial aspect of the upper arm.

Point: Chize (L5)
Location: On the cubital crease, it is near the radial border of the tendon of m. biceps brachii, locating is made with the elbow slightly flexed.
Indications: Cough, asthma, hemoptysis, sore throat, pain and swelling of the medial aspect of the elbow joint.

Point: Kongzui (L6)
Location: 7 cun above the transverse wrist crease, on the line between Pt. Chize and Pt. Taiyuan.
Indications: Cough, headache, pain in the chest, stiffness and severe pain in the neck, asthma, pain and disability of movements of the elbow and the arm.

Point: Lieque (L7)
Location: 1.5 cun above the transverse wrist crease, above the processus styloideus of the radius.
Indications: Headache, cough, nasal obstruction, facial paralysis.

Point: Jingqu (L8)
Location: 1 cun above Pt. Taiyuan on the radial side of the a. radialis.
Indications: Cough, dyspnea, sore throat, pain in the wrist and hand.

Point: Taiyuan (L9)
Location: On the transverse wrist crease, in the depression on the radial side of a. radialis.
Indications: Acrotism; asthma; pain in the chest, back and shoulder; diseases of the wrist and its surrounding soft tissues.

Point: Yuji (L10)
Location: In the middle of the palmar surface of the os metacarpal I at the junction of the red and white skin.
Indications: Cough asthma, fever sore throat, diseases of the tendon and the synovial sheath of the wrist and hand.

Point: Shaoshang (L11)
Locations: On the radial side of the pollex, 0.1 cun proximal to the corner of the nail.
Indications: Sore throat, fever, coma, respiratory failure.

CLASSICAL APs PERTAINING TO THE LARGE INTESTINE MERIDIAN (LI)

Point: Shangyang (LI1)
Location: On the tip of the radial side of the index, 0.1 cun proximal to the corner of the nail.
Indications: Fever, sore throat.

Point: Erjian (LI2)
Location: In the depression anterior to the radial side of the second articulation of metacarpo-phalangeal joint.
Indications: Dizziness, epistaxis, toothache, sore throat.

Point: Sanjian (LI3)

CLASSICAL APs PERTAINING TO THE LARGE INTESTINE MERIDIAN (LI) *(cont'd)*

Location: In the depression on the radial side of the index finger posterior to the small head of the os metacarpal II, half-clenched fist for locating this point.
Indications: Ophthalmalgia, lower toothache, trigeminal neuralgia, sore throat, redness and swelling of the dorsum of the hand.
Point: Hegu (LI4)
Location: On the middle point of the os metacarpal II, on the prominence of the first m. inter ossei dorsales slightly towards the side of the index.
Indiactions: Common cold, facial paralysis, hemiphlegia, neurasthenia, toothache and various kinds of pain.
Point: Yangxi (LI5)
Location: On the radial end of the dorsal crease of the wrist, when the thumb is tilted upward, it is in the depression between tendons of the m. extensor pollicis longus and brevis m.
Indications: Headache, redness of the eye, deafness, tinnitus, laryngitis, pain of the wrist, delinum, anxiety, indigestion in infants and children.
Point: Pianli (LI6)
Location: 3 cun above Pt. Yangxi, in the lateral depression of the radius.
Indication: Tonsillitis, facial paralysis, forearm neuralgia, uropnea, edema, epistaxis, etc.
Point: Wenlui (LI7)
Locations: 5 cun above Pt. Yangxi.
Indication: Headache, sore throat, borborygmus, abdominal pain, pain of the shoulder and back regions.
Point: Xialian: (LI8)
Locations: 4 cun below Pt. Quchi.
Indication: Headache, dizziness, abdominal pain, pain in the elbow and arm, indigestion.
Point: Shanglian (LI9)
Location: 3 cun below Pt. Quchi.
Indications: Hemiplegia, numbness of the foot and hand, sprain, borborygmus, abdominal pain.
Point: Shousanli (LI10)
Location: 2 cun below Pt Quchi.
Indications: Hemiplegia, parotitis, rheumatic neuralgia of the elbow and arm, facial paralysis, headache, ophthalmalgia, deafness.
Point: Quchi (LI11)
Location: Between the end of the cubital crease and the epicondylus lateralis of the humerus, forming a 90 degree angle to locate the point.
Indications: Hemiphlegia, joint pain of the upper extremities, hypertension, high fever, measles, pain of the back.

Point: Zhouliao (LI12)
Location: 1 cun above Pt. Quchi.
Indications: Pain in the elbow and shoulder regions, numbness, pain of the elbow joint.
Point: Hand- Wuli (LI13)
Location: 3 cun above Pt. Quchi.
Indications: Hemoptysis, tuberculosis of the cervicle lymph nodes, pneumonia, pleurisy, pain in the elbow and arm.
Point: Binao (LI14)
Location: On the lateral aspect of the upper arm, slightly anterior the insertion of m. deltoideus on the line between Pt. Tianyu and Pt. Quchi.
Indications: Pain in the shoulder and arm, paralysis of the upper extremity, eye diseases.
Point: Jianyu (LI15)
Location: When the arm is in abduction at 90 degrees, there are two depressions upon the shoulder joint, the point is in the depression between acromion and great tubercle of the humerus.
Indications: Pain in the shoulder, arm and the joints of upper extremity, hemiplegia, paralysis, hypertension, hidrosis, etc.
Point: Jugu (LI16)
Location: In the depression between the acromial end of the clavicle and the spine of the scapula.
Indications: Diseases of shoulder joint and soft tissue, hemoptysis, tuberculosis of the cervical lymph nodes.
Point: Tianding (LI17)
Location: 1 cun below Pt. Futu, at the posterior border of the m. sterno cleido mastoideus.
Indications: Tonsillitis, tuberculosis of cervical lymph nodes.
Point: Futu (LI18)
Location: On the posterior border of the m. sterno-cleido-mastoideus, at the level of the prominentia laryngea.
Indications: Cough, excessive sputum, sore throat.
Point: Halio (LI19)
Location: 0.5 cun lateral to Pt. Renzhong just below lateral border if the naris.
Indications: Epistaxis, nasal obstruction, facial paralysis.
Point: Yingxiang (LI20)
Location: Between the nasolabial groove and the midpoint of the lateral border of the alae nasi
Indications: Diseases of the nasal cavity, facial paralysis, trigeminal neuralgia, ascariasis of the bile duct.

CLASSICAL APs PERTAINING TO THE STOMACH MERIDIAN (S)

Point: Chengqi (S1)
Location: When the patient is looking straight forward, it is just below the pupilla above the margin of infra- orbitalis. When the patient closes the eyes and lies on his back, it is inferior to the eyeball, locating the point along the margin of the orbit.
Indications: Myopia, redness and swelling of the eye, night blindness, spasm of eyelids optic atrophy, etc.
Point: Sibai (S2)
Location: When the eyes look straight forwards, 1 cun below the pupilla at the foramen infra-orbitale.
Indications: Facial paralysis, spasm of facial muscles, trigeminal neuralgia, conjunctivitis, myopia, frequently used in acupuncture anaesthesia in ophthalmological operations.
Point: Juliao (S3)
Location: When the eyes look straight forwards, it is inferior to the pupilla at the level of the lower border of the alae nasi.
Indications: Facial paralysis, facial spasm, trigeminal neuralgia, stuffiness nose, epistaxis.
Point: Dicang (S4)
Location: 0.4 cun lateral to the mouth angle.
Indications: Facial paralysis, salivation, trigeminal neuralgia.
Point: Daying (S5)
Location: 0.5 cun below Pt. Jiache, a groove appears when the mouth is closed and the cheeks are blown.
Indications: Parotiditis, lock jaw, facial paralysis, toothache, etc.
Point: Jiache (S6)
Location: One finger width anterior and superior to the angle of lower jaw, at the prominence of the m. masseter during mastication.
Indications: Trigeminal neuralgia, toothache, parotiditis, facial paralysis, lock jaw, stiffness and the pain of the neck.
Point: Xiaguam (S7)
Location: In the depression formed by arcus zygomaticus and incisura mandibulae.
Indications: Toothache, arthritis of the mandible, trigeminal neuralgia, deafness.
Point: Touwei (S8)
Location: 0.5 cun superior to the angle between two hairlines at the front.
Indications: Headache, dizziness and vertigo.
Point: Renying (S9)
Location: 1.5 cun lateral of the prominentia larygea, on the anterior border of m. sterno-cleido mastoideus, at the pulsation of a. carotis communis (avoid puncturing the blood vessel)
Indications: Hypertension, asthma, sore throat, hemoptysis, goiter, dysphonia, etc.
Point: Shuitu (S10)
Location: Between Pt. Renying and Pt. Qishe at the anterior border of the m. sterno-cleido mastoideus.
Indications: Sore throat, cough, short of breath.
Point: Qishe (S11)
Location: On the upper border of the medial end of the clavicle, between the sternal head and the clavicular head of the m. sterno-cleido mastoideus.
Indications: Sore throat, asthma, goiter, tuberculosis of the cervicle lympth nodes.
Point: Quepen (S12)
Location: In the centre of the supra- clavicular fossa, 4 cun lateral to Pt. Tiantu.
Indications: Cough, asthma, hiccup, tuberculosis, of the cervicle lymph nodes, intercostals neuralgia, etc.
Point: Qihu (S13)
Location: Inferior to the lower border of the middle point of the clavicula, 4 cun lateral to Pt. Xuanji
Indications: Bronchitis, asthma, hiccup, intercostal neuralgia.
Point: Kufang (S14)
Location: In the first intercostal space of the midclavicular line, take the point while lying on back.
Indications: Hiccup, chest pain fullness in chest and costal region.
Point: Wuyi (S15)
Location: On the nipple line at the level of the second intercostal space.
Indications: Cough, dyspnea, fullness and pain of the chest and costal region, mastitis.
Point: Yingchuang (S16)
Location: On the nipple line at the level of the third intercostal space.
Indications: Cough, dyspnea fullness and pain of the chest and costal region, mastitis.
Point: Ruzhong (S17)
Location: In the centre of the papilla mammae, on the mid-clavicular line at the level of the fourth intercostal space.
Indications: As the land mark for locating the points on the abdomen and chest.
Point: Rugen (S18)

CLASSICAL APs PERTAINING TO THE STOMACH MERIDIAN (S) *(cont'd)*

Location: Directly below the papilla mammae, in the fifth intercostal space; take the point while lying on back.
Indications: Mastitis, short of milk.
Point: Burong (S19)

Location: 6 cun above the umbilicus, 2 cun lateral to Ren Mai.
Indications: Stomach ache, vomiting, abdominal distension, loss of appetite.
Point: Chengman (S20)

Location: 5 cun above the umbilicus, 2 cun lateral to Pt. Shangwan.
Indications: Stomach-ache, vomiting, borborygmus, hernia pain, indigestion.
Point: Liangmen (S21)

Location: 2 cun lateral to Pt. Zhongwan.
Indications: Gastric and duodenal ulcers, acute and chronic gastritis, gastric spasm, gastric neurosis, etc.
Point: Guanmen (S22)

Location: 3 cun above umbilicus, 2 cun lateral to Ren Mai.
Indications: Abdominal distension, loss of appetite, borborygmus, diarrhea, edema, etc.
Point: Taiyi (S23)

Location: 2 cun above umbilicus 2 cun lateral to Pt. Xiawan.
Indications: Stomach ache, hernia, Hongkong foot, enuresis, neurosis and psychosis.
Point: Huaroumen (S24)

Location: 1 cun above umbilicus, 2 cun lateral to Ren Mai.
Indications: Stomach ache, vomiting, psychosis, etc.
Point: Tianshu (S25)

Location: 2 cun lateral to the umbilicus.
Indications: Diarrhea, bacillary dysentery, enteritis, gastritis, intestinal ascariasis, appendicitis, constipation of infants, etc.
Point: Wailing (S26)

Location: 1 cun below umbilicus, 2 cun lateral to Ren Mai.
Indications: Abdominal pain, hernia dysmenorrheal, etc.
Point: Daju (S27)

Location: 2 cun below umbilicus, 2 cun lateral to Ren Mai.
Indications: Fullness in lower abdomen, dysuria, hernia, nocturnal emission, ejaculation, praecox.
Point: Shuidao (S28)

Location: 3 cun below umbilicus, 2 cun lateral to Pt. Guanyuan.
Indications: Fullness in lower abdomen, hernia, dysuria, dysmenorrhea.
Point: Guilai (S29)

Location: 2 cun lateral to Pt. Zhongji.
Indications: Irregular menses, dysmenorrhea, chronic inflammatory disease of pelvis, adnexitis, endometritis, prolapse of uterus, impotence, hernia, etc.
Point: Qichong (S30)

Location: Latero-superior to the tuberculum pubicum, 2 cun lateral to the midline, upper portion of inguinal region, medical to the artery.
Indications: Genital diseases of both female and male, hernia, etc.
Point: Biguan (S31)

Location: At the junction of a line between spina iliac anterior-superior and lateral upper border of the patella and horizontal line of the perineum.
Indications: Numbness of lower extremity, paralysis, inguinal lymphadenitis, arthritis of the knee joint, lumbago, etc.
Point: Femur-Futu (S32)

Location: 6 cun above the superior border of the patella, on the line between the spina iliac anterior superior and latero-superior border of the patella.
Indications: Paralysis of lower extremity, numbness, arthritis of the knee joint, ulticaria.
Point: Yinshi (S33)

Location: 3 cun above the latero-superior border of the patella, between m. rectus femoris and m. lateralis.
Indications: Arthritis of the knee joint, paralysis of lower extremity, etc.
Point: Liangqiu (S34)

Location: 2 cun above supra-lateral border of the patella.
Indications: Stomach ache mastitis, gastritis and diarrhea, diseases of knee joint and the anterior aspect of the leg.
Point: Dubi (S35)

Location: In the depression at the lower border of the patella and lateral to the lig. patella.
Indications: Diseases of knee joint and its surrounding soft tissues.
Point: Zusanli (S36)

Continued

CLASSICAL APs PERTAINING TO THE STOMACH MERIDIAN (S) *(cont'd)*

Location: 6 cun below Pt. Dubi, one finger width lateral to the crista anterior tibiae.
Indications: Gastritis, peptic ulers, enteritis, acute pancreatitis, indigestion of infants, diarrhea, dysentery, insomnia, hypertension, etc.
Point: Shangjuxu (S37)
Location: 3 cun below Pt. Zusanli.
Indications: Appendicitis, dysentery, diarrhea and diseases of lower extremity.
Point: Tiaokou (S38)
Location: 8 cun above lateral condyle of the ankle, one finger width lateral to the crista anterior tibiae.
Indications: Arthritis of the knee joint, paralysis of the lower extremity, sciatic neuralgia.
Point: Xiajuxu (S39)
Location: 1 cun below Pt. Tiaokou.
Indications: Enteritis, paralysis of the lower extremity, intercostal neuralgia orchitis with pain referred to lower abdomen.
Point: Fenglong (S40)
Location: 1 cun lateral to Pt. Tiaokou.
Indications: Cough, excessive sputum, hemiplegia, sore throat, constipation, dizziness, mania depressive psychosis, etc.
Point: Jiexi (S41)
Location: In the centre of dorsal crease of ankle joint, between tendon of the m. extensor hallucis longus and tendon of the m. extensor digitorum longus.
Indications: Headache, drop foot, diseases of the lower extremity and soft tissue around ankle joint.
Point: Chongyang (S42)
Location: Anterior inferior to Pt. Jiexi, at the highest spot of dorsum of foot.
Indications: Pain in dorsum of foot, paralysis of lower extremity, toothache epilepsy.
Point: Xiangu (S43)
Location: Between the dorsum of the os metatarsale II and III, in the depression posterior to the art. Metatarsophalangeae.
Indications: Facial swelling, edema, borborygmus, abdominal pain, swelling and pain in the dorsum of the foot.
Point: Neiting (S44)
Location: Proximal to the web margin between the 2nd and 3rd toes.
Indications: Toothache, trigeminal neuralgia, tonsillitis, pain in epigastrium.
Point: Lidui (S45)
Location: Lateral to the nail of the second toe, 0.1 cun distal to the corner of the nail.
Indications: Anemia neurasthenia, tonsillitis, indigestion, hysteria, etc.

CLASSICAL APs PERTAINING TO THE SPLEEN AND PANCREAS MERIDIAN (SP)

Point: Yinbai (SP1)
Location: On the medial aspect of the hallux, 0.1 cun proximal to the corner of nail.
Indications: Abdominal distension, menometrorrhagia, dream-disturbed sleep, convulsion, mental disorder.
Point: Dadu (SP2)
Location: On the medial aspect of the hallux, antero-inferior to the first art. metatarso-phalanx at the junction of red and white skin.
Indications: Gastric pain, abdominal distension, indigestion, nausea and vomiting, diarrhea, febrile diseases with hypohydrosis.
Point: Taibai (SP3)
Location: On the medial aspect of the hallux, postero-inferior to the small head of os metatarsale I.
Indications: Gastric pain, abdominal distension, lassitude, dysentery, etc.
Point: Gongsun (SP4)
Location: On the medial aspect of the foot, in a depression at medio-inferior border of the os metatarsale I and at the junction of the red and white skin.
Indications: Gastric pain, vomiting, indigestion, diarrhea, menorrhalgia.
Point: Shangqui (SP5)

CLASSICAL APs PERTAINING TO THE SPLEEN AND PANCREAS MERIDIAN (SP) *(cont'd)*

Location: In the depression at the anterior-inferior border to the maleolus medialis.
Indications: Borborygmus, abdominal distension, constipation, jaundice, diarrhea, indigestion, pain of the malleolus region.
Point: Sanyinjiao (SP6)
Location: 3 cun above the highest point of the malleolus medialis at the posterior border of the tibia.
Indications: Irregular menstruation, menorrhagia nocturnal emission, impotence, abdominal pain, enuresis, diarrhea, hemiplegia, neurasthenia.
Point: Lougu (SP7)
Location: 6 cun above the highest point of the malleolus medialis.
Indications: Abdominal distension, borborygmus, numbness of the leg and knee.
Point: Diji (SP8)
Location: 3 cun below Pt. Yinlingquan.
Indications: Irregular menstruation, menorrhagia dysentery, abdominal distension.
Point: Yinlingquan (SP9)
Location: In the depression of the lower border of the condylus medialis of the tibia, when the knee is flexed.
Indications: Abdominal pain, edema, dysuria, enuresis, emission, irregular menstruation, dysentery.
Point: Xuehai (SP10)
Location: 2 cun above the antero-superior border of the patella, when the knee is flexed.
Indications: Irregular menstruation, functional uterine bleeding, urticaria, menorrhagia.
Point: Jimen (SP11)
Location: 6 cun above Pt. Xuehai.
Indications: Urethritis, incontinence of urine, lymphadentis of inguinal region.
Point: Chongmen (SP12)
Location: 3.5 cun lateral to the symphysis pubis, on the lateral side of the a. femoralis.
Indications: Abdominal pain, hernia, painful hemorrhoids, dysuria.
Point: Fushe (SP13)

Location: 0.7 cun above Pt. Chongmen.
Indications: Abdominal pain, hernia, splenomegalia.
Point: Fujie (SP14)
Location: 1.3 cun above Pt. Daheng.
Indications: Peri-umbilical pain, hernia, diarrhea due to abdominal cold.
Point: Daheng (SP15)
Location: 4 cun lateral to umbilicus.
Indications: Abdominal distension, diarrhea, constipation, intestinal paralysis, intestinal parasitic diseases.
Point: Fuai (SP16)
Location: 3 cun above Pt. Daheng.
Indications: Abdominal pain, indigestion, constipation, dysentery.
Point: Shiddou (SP17)
Location: In the fifth intercostal space, 6 cun lateral to Ren Mai.
Indications: Pain and distension of lower chest and hypochondrium.
Point: Tianxi (SP18)
Location: In the fourth intercostal space, 6 cun lateral to Ren Mai.
Indications: Pain in the chest, thoracalgia, cough, mastitis, obligalactia.
Point: Xiongxiang (SP19)
Location: In the third intercostal space, 6 cun lateral to Ren Mai.
Indications: Pain and distension of lower chest and hypochondrium.
Point: Zhourong (SP20)
Location: In the second intercostal space, 6 cun lateral to Ren Mai.
Indications: Distension of lower chest and hypochondrium, cough, pain of hypochondrium.
Point: Dabao (SP21)
Location: In the sixth intercostal space, on the midaxillary line.
Indications: Pain of lower chest and hypochondrium, general aching, weakness of limbs.

CLASSICAL APs PERTAINING TO THE HEART MERIDIAN (H)

Point: Jiquan (H1)
Location: At the centre of the fossa axillaries on the medial side of the a. axillaries when the arm is abducted.
Indications: Arthritis of the shoulder, pain in the lower chest and hypochondriac region.
Point: Qingling (H2)
Location: 3 cun above Pt. Shaohai.
Indications: Pain of hypochondrium, yellow discolouration of the eye ball, pain of shoulder and back.
Point: Shaohai (H3)
Location: Between the ulnar end of the cubital crease and the epicondylus medialis of the humerus.
Indications: Diseases of the elbow joint and the palmar side of the forearm, neurasthenia, schizophrenia, intercostal neuralgia.
Point: Lingdao (H4)
Location: 1.5 cun above Pt. Shenmen, on the ulnar border of dorsal surface of the hand.
Indications: Cardiac pain, mental disorder, diseases of ulnar aspect.
Point: Tongli (H5)
Location: 1 cun above Pt. Shenmen, on the ulnar side of the tendon of m. flexor carpi ulnaris.
Indications: Palpitation, angina pectoris, aphasia due to hysteria, pain of the wrist and arm, neurasthenia.
Point: Yinxi (H6)
Location: 0.5 cun below Pt. Tongli.
Indications: Angina pectoris, cardiac arrhythmia, night sweating.
Point: Shenmen (H7)
Location: When the forearm is in supination, it is in the depression at ulnar end of the wrist crease, radial to the tendon of m. flexor carpi ulnaris.
Indications: Amnesia, insomnia, dreaminess, angina pectoris, hysteria.
Point: Shaofu (H8)
Location: On the ulnar side of Pt. Laogong, between the ossa metacarpale IV and V.
Indications: Cardiac arrhythmia, angina pectoris, toothache.
Point: Shaochong (H9)
Location: On the radial aspect of the digitus minimus, about 0.1 cun proximal from the corner of the nail.
Indications: Coma, insanity, angina pectoris.

CLASSICAL APs PERTAINING TO THE SMALL INTESTINE MERIDIAN (SI)

Point: Shaoze (SI1)
Location: On the ulnar aspect of the digitus minimus, 0.1 cun proximal from the corner of the nail.
Indications: Headache, mastitis, difficiency of lactation, pterygium.
Point: Qiangu (SI2)
Location: Anterior to the ulnar side of the fifth art. metacarpophlangae when clenching a fist, a transverse crease is formed there and distal to which the point stands.
Indications: Numbness of the finger, leukoma, tinnitus, mastitis.
Point: Houxi (SI3)
Location: On the ulnar side of caput os matacarpale V; when the fist is clenched halfway it is in the depression at the end of palmar transverse crease.
Indications: Tinnitus, deafness, epilepsy, malaria, pain in the shoulder and the back, parietal headache, lumbago, intercostal neuralgia, acute sprain of the lumbar region.
Point: Hand Wanggu (SI4)
Location: On the ulnar side of the back of the hand, in the depression amid the base of the os metacarpale V the os hamatum and the pisiforme.
Indications: Athritis of the wrist, elbow and phalangeal joints, headache, tinnitus, vomiting cholecystitis.
Point: Yanggu (SI5)

CLASSICAL APs PERTAINING TO THE SMALL INTESTINE MERIDIAN (SI) *(cont'd)*

Location: In the depression of the ulnar end of the transverse crease of back of the wrist, between processus styloideus of the ulna and the os triquetrum.
Indications: Pain in the wrist, parositis, mental disease, deafness, tinnitus, etc.
Point: Yanglao (SI6)

Location: Flex the elbow with palm against the chest, the point is on the bony cleft on the radial side of the processus styloideus of the ulna.
Indications: Joint pain of the upper extremity, pain in the shoulder and the back, hemiplegia, lumbago, wryneck, blurring vision.
Point: Zhizheng (SI7)

Location: 5 cun above the ulnar end of transverse crease of the back of the wrist, on the line between Pt. Yanggu and Pt. Xiaohai.
Indications: Rigidity of neck, pain in the elbow, arm and fingers, mental diseases.
Point: Xiao Hai (SI8)

Location: In sulcus n. ulnaris between the olecranon of the ulna and the epicondylus medialis of the humerus, flex the elbow when locating the point.
Indications: Pain in the neck, pain in the shoulder and the back, pain in the elbow joints, epilepsy, diseases of ulnar side of the upper extremity.
Point: Jianzhen (SI9)

Location: With hands close to thighs 1 cun above the posterior anxillary fold.
Indications: Diseases of the shoulder joint and its surrounding soft tissues, paralysis of the upper extremity, anxillary hidrosis, etc.
Point: Naoshu (SI10)

Location: Superior and slight lateral to the Pt. Jianzhen, at the lower border of the spine scalpulae.
Indications: Apoplexy and hemiplegia, hypertension, pain in the shoulder joint, also used in the abduction weakness of the arm.
Point: Tianzong (SI11)

Location: In the centre of the fossa infraspinata of the scapula.
Indications: Pain in the scapular region, pain in poster-lateral aspect of the elbow and arm, asthma, deficiency of lactation.
Point: Bingfeng (SI12)

Location: In the centre of the fossa supraspinata of the scapula, directly above Pt. Tianzong, in the depression when the arm is lifted.
Indications: Pain in the scapular region with difficulty in lifting the arm, ache or numbness of the upper extremity.
Point: Quyuan (SI13)

Location: On the medial end of the fossa supraspinate of the scapula, at the level of processus spinosus of the third vertebra thoracica.
Indications: Diseases of the scapular region.
Point: Jianwaishu (SI14)

Location: 3 cun lateral to the lower border of the processus spinosus of the first vertebra thoracic.
Indications: Diseases of the neck, back and shoulder.
Point: Jianzhongshu (SI15)

Location: 2 cun lateral to the lower border of the seventh vertebra cervicalis.
Indications: Diseases of the shoulder and the back, cough, asthma.
Point: Tianchuang (SI16)

Location: 3.5 cun lateral to laryngeal cartilage, on the posterior border of the m. sterno-cleidomastoideus, 0.5 cun posterior to the Pt. Neck-Futu.
Indications: Sore throat, goiter, tinnitus, deafness, stiffness and pain of the neck, etc.
Point: Tianrong (SI17)

Location: Posterior to the angle of lower jaw, in the depression on the anterior border of m. sternocleido-mastoideus.
Indications: Tonsillitis, pharyngitis, painful swelling of the neck, asthma, etc.
Point: Quanliao (SI18)

Location: In the centre of the lower border of the os zygomaticum at the level Pt. Yingxiang, directly below the lateral canthus of the eye.
Indications: Trigeminal neuralgia, facial spasm and facial paralysis, etc.
Point: Tinggong (SI19)

Location: Anterior to helix in the depression posterior to the art, temporomandilaris when opening the mouth.
Indications: Tinnitus, deafness, otitis media, toothache, facial paralysis, deaf-mutism.

CLASSICAL APs PERTAINING TO THE URINARY BLADDER MERIDIAN (B)

Point: Jingming (B1)
Location: In the depression medial and superior to the medial canthus.
Indications: Conjunctivitis, strabismus, myopia, glaucoma, optic neuritis, retinitis, optic atrophy, etc.

Point: Zanzhu (B2)
Location: In the medial end of the eyebrow, above the medial canthus.
Indications: Headache, trigamnial neuralgia, facial paralysis, glaucoma.

Point: Meichong (B3)
Location: 0.5 cun inside the hairline between Pt. Shengting and Pt. Quchi.
Indications: Headache, nasal obstruction, dizziness epilepsy, etc.

Point: Quchai (B4)
Location: 0.5 cun inside the hairline at the junction between lateral 2/3 and medial 1/3 of the line between Pt. Shengting and Pt. Touwei.
Indications: Headache, nasal obstruction, epistaxis, eye diseases, etc.

Point: Wuchu (B5)
Location: 1 cun inside the hairline above Pt. Quchai.
Indications: Headache, dizziness, rhinitis, epilepsy, etc.

Point: Chengguang (B6)
Location: 1.5 cun posterior to Pt. Wuchu.
Indications: Headache, common cold, leucoma, rhinitis, dizziness, etc.

Point: Tongtian (B7)
Location: 1.5 cun posterior to Pt. Chengguang.
Indications: Headache, dizziness, nasal obstruction, epistaxis, sinusitis.

Point: Luogue (B8)
Location: 1.5 cun posterior to Pt. Tongtin.
Indications: Dizziness, facial paralysis, rhinitis, goiter, vomiting, etc.

Point: Yuzhen (B9)
Location: On the lateral side of the superior border of the protuberantia occipitalis externa, above Pt. Tianzhu.
Indications: Headache, dizziness, myopia, etc.

Point: Tianzhu (B10)
Location: 1.3 cun lateral to Pt. Yamen.
Indications: Headache, stiffness of the neck, sore throat.

Point: Dashu (B11)
Location: 1.5 cun Lateral to Pt. Taodao below the processus spinosus of the first vertebra thoracicae.
Indications: Fever, cough, headache, pain of the shoulder, stiffness of the neck.

Point: Fengmen (B12)
Location: 1.5 cun lateral to the processus spinosus of the second vertebra thoracica.
Indications: Common cold, cough, fever, headache, asthma, chronic rhinitis, diseases of the back. This point is frequently used for acupuncture anesthesia in head and brain surgery.

Point: Feishu (B13)
Location: 1.5 cun lateral to the lower border of the processus spinosus of the third vertebra thoracica.
Indications: Common cold, nasal obstruction, cough, asthma, night sweating, diseases of the back.

Point: Jueyinshu (B14)
Location: 1.5 cun lateral to the lower border of the processus spinosus of the fourth vertebra thoracica.
Indications: Angina pectoris, arrhythmia tachycardia and other heart diseases, epilepsy, mental disorder, insomnia, pain in the chest.

Point: Xinshu (B15)
Location: 1.5 cun lateral to the lower border of the processus spinosus of the fifth vertebra thoracica.
Indications: Palpitation, distress, cough, weak memory, angina pectoris, arrthythmia tachycardia, neurasthenia, etc.

Point: Dushu (B16)
Location: 1.5 cun lateral to the lower border of the processus spinosus of the sixth vertebra thoracica.
Indications: endocarditis, abdominal pain, borborygmus, spasm of the diaphragm, mastitis, psoriasis, etc.

Point: Geshu (B17)
Location: 1.5 cun lateral to the lower border of the processus spinosus of the seventh vertebra thoracica (Pt. Zhiyang).
Indications: Chronic hemorrhagic disease, anemia, acute infection of the bile tract, belching, spasm of oesophagus, cough, asthma, pulmonary tuberculosis, etc.

Point: Ganshu (B18)

CLASSICAL APs PERTAINING TO THE URINARY BLADDER MERIDIAN (B) *(cont'd)*

Location: 1.5 cun lateral to the lower border of the processus spinosus of the ninth vertebra thoracica.
Indications: Diseases of the liver and gall bladder, jaundice, pain of the lower chest, gastric diseases, hemoptysis, epistaxis, redness of the eye, night blindness, glaucoma, back pain.
Point: Danshu (B19)
Location: 1.5 cun lateral to the lower border of the processus spinosus of the tenth vertebra thoracica.
Indications: Jaundice, bitterness in the mouth, pain of the lower chest, fever and sweating due to tuberculosis, diseases of the back.
Point: Pishu (B20)
Location: 1.5 cun lateral to the lower border of the processus spinosus of the eleventh vertebra thoracica.
Indications: Abdominal distension, jaundice, vomiting, diarrhea, dysentery, edema, weakness and dysfunction of the stomach and spleen, indigestion, hepatitis, back pain, etc.
Point: Weishu (B21)
Location: 1.5 cun lateral to the lower border of the processus spinosus of the twelfth vertebra thoracica.
Indications: Pain of the lower chest, pain of epigastrium, abdominal distension, regurgitation, borborygmus, weakness and dysfunction of the stomach and spleen, indigestion, chronic diarrhea.
Point: Sanjiaoshu (B22)
Location: 1.5 cun lateral to the lower border of the processus spinosus of the first vertebra lumbales.
Indications: Abdominal distension, vomiting, diarrhea, dysentery, edema, infection of the urinary tract, back pain.
Point: Shenshu (B23)
Location: 1.5 cun lateral to the lower border of the processus spinosus of the second vertebra lumbales.
Indications: Infection of the urinary tract, impotence nocturnal emission, irregular menstruation, leucorrhea, retention of urine, dysfunction, of urinary tract, asthma, tinnitus, deafness, chronic diarrhea, lumbago.
Point: Qihaishu (B24)
Location: 1.5 cun lateral to the lower border of the processus spinosus of the third vertebra lumbales.
Indications: Abdominal pain and distension, borborygmus, constipation, lumbago, etc.
Point: Dachangshu (B25)
Location: Below processus spinosus of the fourth vertebra lumbales, 1.5 cun lateral to Pt. Yaoyangguan.
Indications: Abdominal pain and distension, borborygmus, constipation, lumbago sciatic neuralgia.
Point: Guanyuanshu (B26)
Location: 1.5 cun lateral to the lower border of the processus spinosus of the 5th vertebra lumbales.
Indications: Abdominal distension, diarrhea, lumbago, nycturia, thirst, frequent urination of dysuria.
Point: Xiaochangshu (B27)
Location: 1.5 cun lateral to the lower border of the first vertebra sacrales.
Indications: Sciatic neuralgia, lumbago, nocturnal emission, urorrhea, enteritis, constipation, inflammatory diseases of the pelvis.
Point: Pangguanshu (B28)
Location: 1.5 cun lateral to the lower border of the second vertebra sacrales.
Indications: Urgency of urination, dysuria, frequent urination, diarrhea, constipation, lumbago, sciatic neuralgia.
Point: Zhonglushu (B29)
Location: At the level of the third posterior sacral foramen, 1.5 cun lateral to the Du Mai.
Indications: Enteritis, lumbago, sciatic neuralgia.
Point: Baihuanshu (B30)
Location: On the lower border of the processus spinosus of the fourth vertebra sacrales, 1.5 cun lateral to Pt. Yaoshu.
Indications: Lumbago, irregular menstruation, leukorrhea, chronic inflammation of pelvic organs, sciatic neuralgia, sacral neuralgia.
Point: Shangliao (B31)
Location: Opposite to the posterior 8 sacralia foramena Pt. Shangliao, Pt. Ciliao, Pt. Zhongliao and Pt. Xialiao are located at the first, second, third, and fourth posterior sacral foramen respectively.
Indications: Lumbago, irregular menstruation, lower abdominal pain, menorrhalgia, leucorrhea, dysuria, impotence, nocturnal emission, prolapse of anus, etc.
Point: Ciliao (B32)

Continued

CLASSICAL APs PERTAINING TO THE URINARY BLADDER MERIDIAN (B) *(cont'd)*

Location: See prior entry
Indications: See prior entry
Point: Zongliao (B33)
Location: See prior entry
Indications: See prior entry
Point: Xialiao (B34)
Location: See prior entry
Indications: See prior entry
Point: Huiyang (B35)
Location: 0.5 cun lateral to the os coccygis.
Indications: Enteritis, hemorrhoids, female genital diseases, impotence, hematuria.
Point: Chengfu (B36)
Location: Midpoint of the gluteal fold.
Indications: Lumbago, sciatic neuralgia, paralysis of the lower extremities, urorrhagia, constipation.
Point: Yinmen (B37)
Location: 6 cun below Pt. Chengfu, on the line between Pt. Chengfu and Pt. Weizhang.
Indications: Lumbago, sciatic neuralgia, paralysis of the lower extremities, paralysis.
Point: Fuxi (B38)
Location: Lateral to poplital fossa, 1 cun above Pt. Weiyang.
Indications: Acute gastroenteritis, cystitis, constipation, paralysis of the lateral aspect of the lower extremities.
Point: Weiyang (B39)
Location: Above the popliteal crease, 1 cun lateral to Pt. Weizhong.
Indications: Nephritis, chyluria, cystitis, constipation, etc.
Point: Weizhong (B40)
Location: Midpoint of the popliteal transverse crease.
Indications: Nocturnal emission, impotence, dysuria, acute lumbago sciatic neuralgia, diseases of the lower extremities and knee joints.
Point: Fufen (B41)
Location: 3 cun lateral to the midline, between the processus spinosus of the second and the third vertebrae thoracicae.
Indications: Pain of the shoulder, neck and back, numbness of the elbow and arm.
Point: Pohu (B42)
Location: 3 cun lateral to the midline, between the processus spinosus of the third and the fourth vertebrae thoracicae.
Indications: Bronchitis, weakness of the chest, asthma, pulmonary tuberculosis, pleurisy, etc.
Point: Gaohuangshu (B43)
Location: 3 cun lateral to the lower border of the processus spinosus of the fourth vertebra thoracica.
Indications: Gastric pain, vomiting, abdominal distension, constipation, lumbago.
Point: Shentang (B44)
Location: 3 cun lateral to the midline, between the processus spinosus of the fifth and sixth vertebrae thoracicae.
Indications: Bronchitis, asthma, intercostal neuralgia, heart diseases, etc.
Point: Yixi (B45)
Location: 3 cun lateral to the midline, between the processus spinosus of the sixth and seventh vertebrae thoracicae.
Indications: Pericarditis, asthma, malaria, intercostal neuralgia, belching, etc.
Point: Geguan (B46)
Location: 3 cun lateral to the midline, between the processus spinosus of the seventh and eighth vertebrae thoracicae.
Indications: Intercostal neuralgia, oesophagael spasm, gastric hemorrhage.
Point: Hunmen (B47)
Location: 3 cun lateral to the midline of the back, between the processus spinosus of the ninth and tenth vertebrae thoracicae.
Indications: Neurothenia, diseases of the liver and gall bladder, pleuritis, gastralgia, etc.
Point: Yanggang (B48)
Location: 3 cun lateral to the midline of the back, between the processus spinosus of the tenth and eleventh vertebrae thoracicae.
Indications: Hepatitis, chole-cystitis, gastritis, etc.
Point: Yishe (B49)
Location: 3 cun lateral to the lower border of the processus spinosus of the eleventh vertebra thoracica.
Indications: Lumbago, abdominal distension, indigestion, thirst, jaundice, etc.
Point: Weicang (B50)
Location: 3 cun lateral to the lower border of the processus spinosus of the twelfth vertebra thoracica.
Indications: Gastralgia, vomiting, abdominal distension, constipation, lumbago.
Point: Huangmen (B51)

CLASSICAL APs PERTAINING TO THE URINARY BLADDER MERIDIAN (B) *(cont'd)*

Location: 3 cun lateral to the midline, between the processus spinosus of the first and second vertebrae lumbales.
Indications: Mastitis, upper abdominal pain, lumbago, paralysis of the lower extremities.
Point: Zhishi (B52)
Location: 3 cun lateral to the midline of the back, between the processus spinosus of the second and third vertebra lumbales.
Indications: Nacturnal emission, impotence, irregular menstruation, enuresis, chronic lumbago, etc.
Point: Baohuang (B53)
Location: 3 cun lateral to the midline of the back, between the processus spinosus of the second and third vertebra sacrales.
Indications: Diseases of the lumbar and sacral regions.
Point: Zhibian (B54)
Location: 3 cun lateral to the midline of the back, on the lower border of the fourth vertecrae sacrales.
Indications: Pain of lumbar and sacral regions, paralysis of the lower extremities, dysuria hemorrhoid, etc.
Point: Heyang (B55)
Location: 2 cun below Pt. Weizhong.
Indications: Lumbago and leg pain, metorrhagia, painful hernia.
Point: Chengjin (B56)
Location: At the midpoint of the curve between Pt. Heyang and Pt. Chengshan.
Indications: Headache, severe pain of lumbar region and back, pain of the leg, paralysis of lower extremities, hemorrhoids, etc.
Point: Chengshan (B57)
Location: At the midpoint of the line between Pt. Weizhong and Pt. Kunlun.
Indications: Pain of back and thigh, sciatic neuralgia, spasm of the m. gastrocnemius, paralysis, hemorrhoids, prolapcusani, etc.
Point: Feiyang (B58)
Location: 7 cun Pt. Kunlun latero-inferior to Pt. Chengshan.
Indications: Rheumatic athritis, nephritis, cystitis, Hongkong foot, hemmorhoid, epilepsy, lumbago, pain of the leg.
Point: Fuyang (B59)
Location: 3 cun above Pt. Kunlun posterior to malleolus lateralis.
Indications: Pain of the neck, diseases of lumbago-sacral region and lower extremities.
Point: Kunlun (B60)
Location: Middle point between the tip of malleolus lateralis and the tendo calcaneus.
Indications: Headache, back pain, lumbago, sciatic neuralgia, paralysis of lower extremities, severe pain of the neck.
Point: Pushen (B61)
Location: Below Pt. Kunlun, in the depression of the calcaneus.
Indications: Lumbago, pain of the ankle, paralysis of the lower extremities, Hongkong foot, etc.
Point: Shenmai (B62)
Location: In the depression at the lower border of the malleolus lateralis.
Indications: Headache, pain of the neck, epilepsy, dementia, diseases of the posterior aspect of lumbo-sacral region and lower extremities.
Point: Jinmen (B63)
Location: Antero-inferior to Pt. Shenmai, in the lateral depression of the os cuboideum.
Indications: Epilepsy, dementia, infantile convulsion, diseases of posterior aspect of the lumbosacral region and lower extremities.
Point: Jinggu (B64)
Location: Postero-inferior to the tuberosity of the fifth os metatarsale.
Indications: Headache, pain of the neck, myocarditis, meningitis, epilepsy, lumbago, pain of the leg, etc.
Point: Shugu (B65)
Location: Postero-inferior to the small head of the fifth os metatarsale.
Indications: Headache, pain of the neck, malaria, leucoma, epilepsy dementia, etc.
Point: Foot-Tonggu (B66)
Location: In the depression anterior and inferior to the fifth art. metatarsophalangeae.
Indications: Headache, dizziness, asthma, epistaxis, dementia, etc.
Point: Zhiyin (B67)
Location: On the lateral side of the tip of the small toe, about 0.1 cun proximal to the corner of the nail.
Indications: Headache, pain of the neck, difficult labour, abdominal presentation.

CLASSICAL APs PERTAINING TO THE KIDNEY MERIDIAN (K)

Point: Yongquan (K1)
Location: In the centre of the sole of the toes, at the junction between anterior 1/3 and posterior 2/3 of the sole (the length of the toe is not included).
Indications: Shock, sunstroke, hypertension, cerebral hemorrhage, infantile convulsions, hysteria, epilepsy.

Point: Rangu (K2)
Location: In the depression at the inferior border of the tuberositas ossis navicularis.
Indications: Infantile convulsions, pruritus of external genitalia, infection of urinary tract, diabetes mellitus.

Point: Taixi (K3)
Location: At the midpoint of the line between the tip of malleolus medialis and archillis tendon.
Indications: Nephritis, cystitis, enuresis, irregular menses, paralysis of lower extremity.

Point: Dazhong (K4)
Location: Inferior and posterior to medial malleolus, in the depression medial to the insertion of the achillis tendon.
Indications: Asthma, malaria, neurosthenia, hysteria, retention of urine, sore throat, heel pain, etc.

Point: Shuiquan (K5)
Location: 1 cun below Pt. Taixi.
Indications: Anemia, prolapse of the uterus, myopia, etc.

Point: Zhaohai (K6)
Location: 1 cun below the top point of malleolus medialis.
Indications: Irregular menses, neuro-asthenia, epilepsy, constipation, pharyngolaryngitis, tonsillitis.

Point: Fuliu (K7)
Location: 2 cun directly above Pt. Taixi.
Indications: Nephritis, orchitis, night sweating, lumbago, infection of urinary tract, edema.

Point: Jiaoxin (K8)
Location: Anterior to Pt. Fului posterior to the medial border of the tibia.
Indications: Irregular menses, menorrhagia, retention of urine, dysentery, constipation, pain in medial aspect of the lower extremity.

Point: Zhubin (K9)
Location: On the line between Pt. Taixi and Pt. Yingu, on the lower end of the medial belly of the m. gastrocnemius.
Indications: menorrhagia, pain of the leg.

Point: Yingu (K10)
Location: At the medial aspect of the popliteal fossa, between the tendons of m. semitendinosus and m. gastrocnemius.
Indications: Infection of urinary tract, retention of urine, nocturnal emission, impotence, menorrhagia, inguinal hernia, diseases of medial aspect of the knee.

Point: Henggu (K11)
Location: On the superior border of symphysis pubis, 0.5 cun lateral to Pt. Qugu.
Indications: Retention of urine, nocturnal emission, pain of the penis.

Point: Dahe (K12)
Location: 1 cun above Pt. Henggu.
Indications: Nocturnal emission, pain of the penis.

Point: Qixue (K13)
Location: 1 cun above Pt. Dahe.
Indications: Irregular menses, leucorrhea, sterility, infection of urinary tract, diarrhea, etc.

Point: Siman (K14)
Location: 1 cun above Pt. Qixue.
Indications: Irregular menses, leucorrhea, sterility, infection of urinary tract, diarrhea, etc.

Point: Abdomen-Zhongzhu (K15)
Location: 1 cun above Pt. Siman.
Indications: Irregular menses, lumbago, abdominal pain, constipation, etc.

Point: Huangshu (K16)
Location: 1 cun above Pt. Abdomen-Zhongzhu, 0.5 cun lateral to the umbilicus.
Indications: Gastric spasm, painful hernia, enteritis, habitual constipation, hiccup, etc.

Point: Shangqu (K17)
Location: 1 cun above Pt. Huangshu.
Indications: Gastralgia, painful hernia, peritonitis, etc.

Point: Shiguan (K18)
Location: 1 cun above Pt. Shangqu.
Indications: Gastralgia, hiccup, constipation, esophageal spasm, etc.

Point: Yindu (K19)
Location: 1 cun above Pt. Shiguan.
Indications: Emphysema, pleuritis, malaria, abdominal distension, abdominal pain.

Point: Abdomen-Tonggu (K20)
Location: 1 cun above Pt. Yindu.
Indications: Neck rigidity, epilepsy, palpitation, intercostal neuralgia, vomiting, diarrhea, etc.

Point: Youmen (K21)

CLASSICAL APs PERTAINING TO THE KIDNEY MERIDIAN (K) *(cont'd)*

Location: 1 cun above Pt. Abdomen-Tonggu.
Indications: Pain and distension of lower chest, epigastric pain, gastric spasm, etc.
Point: Bulang (K22)
Location: 2 cun lateral to Ren Mai in the fifth intercostal space.
Indications: Pleuritis, intercostal neuralgia, rhinitis, gastritis, bronchitis.
Point: Shengfeng (K23)
Location: 2 cun lateral to Ren Mai in the fourth intercostal space.
Indications: Pleuritis, bronchitis, mastitis, intercostal neuralgia, etc.
Point: Lingxu (K24)
Location: 2 cun lateral to Ren Mai in the third intercostal space.
Indications: Intercostal neuralgia, bronchitis, vomiting, mastitis, etc.
Point: Shencang (K25)
Location: 2 cun lateral to Ren Mai in the second intercostal space.
Indications: Bronchitis, vomiting, intercostal neuralgia.
Point: Yuzhong (K26)
Location: 2 cun lateral to Ren Mai in the first intercostal space.
Indications: Same as Pt. Shencang.
Point: Shufu (K27)
Location: In the lower border of the clavicula, 2 cun lateral to Ren Mai.
Indications: Bronchitis asthma, chest pain, vomiting, abdominal distension, etc.

CLASSICAL APs PERTAINING TO THE PERICARDIUM MERIDIAN (P)

Point: Tianchi (P1)
Location: In the fourth intercostal space, 1 cun lateral to the nipple (for female, it is in the fourth intercostal space, 1 cun lateral to the mid-clavicular line).
Indications: Fullness in chest, pain in lower chest.
Point: Tianquan (P2)
Location: 2 cun below the anterior axillary fold, between the two heads of m. biceps brachii.
Indications: Cough, pain in lower chest, pain of the back and medial aspect of the upper arm.
Point: Quze (P3)
Location: In the transverse cubital crease, at the ulnar side of the tendon of m. biceps brachii, with slight flexion of the elbow for locating.
Indications: Palpitation, angina pectoris, pain in the arm and elbow, tremor of hands, vomiting and diarrhea in acute gastroenteritis.
Point: Ximen (P4)
Location: 5 cun above the transverse crease of wrist, between the tendons of m. palmaris longus and m. flexor carpi radialis.
Indications: Angina pectoris, tachycardia, pleuritis, mastitis, etc.
Point: Jianshi (P5)
Location: 1 cun above Pt. Neiguan between the tendons of m. palmaris longus and m. flexor carpi radiallis.
Indications: Rheumatic heart disease, gastralgia, malaria, hysteria, epilepsy, schizophrenia, etc.
Point: Neiguan (P6)
Location: 2 cun above the transverse crease of the wrist between the tendons of m. palmaris longus and m. flexor carpi radialiss.
Indications: Pain in lower chest, gastralgia, shock, nausea, vomiting, sore throat, hysteria, cardiac arrhythmia.
Point: Daling (P7)
Location: At the midpoint of the transverse crease of wrist between the tendons of m. palmaris longus and m. flexor carpi radialiss.
Indications: Tachycardia, mental disease, intercostal neuralgia, disorders of tendinous sheath of the wrist.
Point: Laogong (P8)
Location: In the middle of the palm, between the middle and the index (between the third and fourth ossa metacarpole) when hand is grasped.
Indications: Mental diseases, epilepsy heatstroke, vomiting, inflammatory disease of the mouth.
Point: Zhong-Chong (P9)
Location: At the tip of the middle finger.
Indications: Coma, fever, heatstroke, angina pectoris, inability of speaking due to rigidity of tongue.

CLASSICAL APs PERTAINING TO THE TRIPLE WARMER MERIDIAN (T)

Point: Guanchong (T1)
Location: On the ulnar side of the ring finger, 0.1 cun proximal to the corner of the nail.
Indications: Sore throat, difficult in speech, conjunctivitis, fever.

Point: Yemen (T2)
Location: In the web between the ring and the little fingers.
Indications: Headache, deafness, malaria, pain in the hand and arm, pain and swelling of fingers, etc.

Point: Hand-Zhongzhu (T3)
Location: Clench the hand when locating, between the ossa metacarpale IV and V, in the depression 1 cun above the art. metacarpophalangeae.
Indications: Deafness, tinnitus, sore throat, disorders of the head, neck, shoulder and back.

Point: Yangchi (T4)
Location: Between the ossa metacarpale III and IV, just above the dorsal crease of wrist, the depression at the ulnar aspect of the tendon of m. extensor digitorum.
Indications: Deafness, malaria, disorders of the wrist joint.

Point: Waiguan (T5)
Location: 2 cun above the transverse crease of dorsum of wrist between the radius and the ulna.
Indications: Common cold, pneumonia, deafness, migraine.

Point: Zhigou (T6)
Location: 1 cun above Pt. Waiguan, between the ulna and the radius.
Indications: Pain of the shoulder and lower chest, constipation, pleuritis, hemiplegia, parotiditis, deafness, tinnitus.

Point: Huizong (T7)
Location: 1 finger breadth lateral to Pt. Zhigou, on the radial side of the ulna.
Indications: Tinnitus, deafness, pain in the upper extremities, epilepsy.

Point: Sanyangluo (T8)
Location: 1 cun above Pt. Zhigou
Indications: Deafness, aphasia, disorders of the forearm.

Point: Sidu (T9)
Location: 5 cun above Pt. Waiguan, between the radius and the ulna.
Indications: Headache, tinnitus, toothache, pain of the forearm, paralysis of upper extremities, neurasthenia, nephritis.

Point: Tainjing (T10)
Location: Superior to the olecranon, in the depression when the elbow is flexed.
Indications: Rigidity of the neck, disorders of the lower chest or upper extremitites.

Point: Qinglengyuan (T11)
Location: 1 cun above Pt. Tianjing.
Indications: Pain in the shoulder, headache, yellowish discolouration of the conjunctiva.

Point: Xiaoluo (T12)
Location: Midway between Pt. Qinglengyuan and Pt. Naohui.
Indications: Headache, stiff neck, pain in the arm, toothache, epilepsy.

Point: Naohui (T13)
Location: At the junction between the line connecting Pt. Jianlao and olecranon and the posterior border of m. deltoideus.
Indications: Pain in the shoulder and arm, hemiplesia, fever and chill.

Point: Jianliao (T14)
Location: Postero-inferior to the acromion, in the depression about 1 cun posterior tp Pt. Jianyu when the arm is raised horizontally.
Indications: Pain in the shoulder and arm, hemiplegia, hypertension, excessive sweating.

Point: Tianliao (T15)
Location: At the superior angle of the scapula, between Pt. Jianjing and Pt. Quyuan.
Indications: Pain in the shoulder and arm, rigidity of the neck.

Point: Tianyou (T16)
Location: Posterior and inferior to the processus mastoideus on the posterior border of m. sternocleido-mastoideus near the hairline.
Indications: Tinnitus, deafness, stiff neck, sore throat.

Point: Yifeng (T17)
Location: Posterior and inferior border of the lobolus auriculae in the depression antero-inferior to the processus mastoideus.
Indications: Tinnitus, deafness, facial paralysis, parotiditis, arthritis of mandible joint, toothache, eye disease, etc.

Point: Qimai (T18)
Location: In the center of the pars mastoideus.
Indications: Deafness, tinnitus.

CLASSICAL APs PERTAINING TO THE TRIPLE WARMER MERIDIAN (T) *(cont'd)*

Point: Luxi (T19)
Location: Midway of the curve between Pt. Qimai and Pt. Jiaosun.
Indications: Headache, tinnitus, ear pain, deafness.
Point: Jiaosun. (T20)
Location: In the hairline above the tip of the auricula.
Indications: Redness and swelling of the ear. toothache, cornea opacity, rigidity of the neck.
Point: Ermen (T21)
Location: Above Pt. Tinggong, in front of the anterior notch of the auricula, in the depression when mouth is open.

Indications: Tinnitus, deafness, toothache, arthritis of mandible point, otitis media, etc.
Point: Ear-Heliao (T22)
Location: Anterior to the upper border of the root of the auricula, posterior to the a. temporalis superficialis.
Indications: Tinnitus, heaviness of the head and headache, facial paralysis, disability of lower jaw.
Point: Sizhukong (T23)
Location: In the depression lateral to the lateral tip of the supercilium.
Indications: Headache, facial paralysis, squint, acute conjunctivitis.

CLASSICAL APs PERTAINING TO THE GALL BLADDER MERIDIAN (G)

Point: Tongziliao (G1)
Location: 0.5 cun lateral to the angulus oculi lateralis.
Indications: Migraine, conjunctivitis, myopia, optic atrophy, and other eye diseases.
Point: Tinghui (G2)
Location: Anterior to the incisura intertragica directly below Pt. Tinggong, in the posterior depression of the articulatio temporomandibularis.
Indications: Tinnitus, deafness, toothache, athritis of the articulatio temporomandibularis.
Point: Shangguan (G3)
Location: Above Pt. Xiaguan, in the depression on the upper margin of the arcus zygomaticus.
Indications: Tinnitus, deafness, otitis media, toothache, lockjaw, facial paralysis, etc.
Point: Hanyan (G4)
Location: On the hairline of the temporal region, between Pt. Touwei and Pt. Xuanlu.
Indications: Migraine, tinnitus, rhinitis, epilepsy convulsion, etc.
Point: Xuanlu (G5)
Location: In the middle of the curve between Pt. Touwei and Pt. Qubin.
Indications: Migraine, pain in lateral canthus, toothache, edema of the face, neurasthenia, etc.
Point: Xuanli (G6)

Location: On the hairline of the temporal region, between Pt. Xuanlu and Pt. Qubin.
Indications: Same as Pt. Xuanlu.
Point: Qubin (G7)
Location: On the hairline in front of the ear apex, one finger width anterior to Pt. Jiaosun.
Indications: Migraine, trigeminal neuralgia, spasm of m. temporalis.
Point: Shuaigu (G8)
Location: Above the ear apex, 1.5 cun within the hairline.
Indications: Migraine, dizziness and vertigo, eye diseases, etc.
Point: Tianchong (G9)
Location: Superior and posterior to the auricula, 0.5 cun posterior to Pt. Shuaigu.
Indications: Toothache, painful swelling of the gum, epilepsy, goitre, etc.
Point: Fubai (G10)
Location: Superior and posterior to the processus mastoideus betwenn Pt. Tianchong and Pt. Hand-Qiaoyin.
Indications: Headache, toothache, deafness, tinnitus, bronchitis, etc.
Point: Head-Qiaoyin (G11)
Location: Posterior to processus masoideus between Pt. Fubai and Pt. Wangu.

Continued

CLASSICAL APs PERTAINING TO THE GALL BLADDER MERIDIAN (G) *(cont'd)*

Indications: Pain in the neck and head, ear pain, deafness, tinnitus, bronchitis, laryngitis, pain in the chest, goitre, etc.
Point: Head-Wangu (G12)
Location: In the depression postero-inferior to processus mastoideus.
Indications: Headache, neck diseases.
Point: Benshen (G13)
Location: On the lateral hairline of the frontal region, on the lateral 1/3 of a line connecting Pt. Shenting and Pt. Touwei.
Indications: Headache, dizziness, stiffness and pain of the neck, pain in the lower chest, epilepsy hemiplegia, etc.
Point: Yangbai (G14)
Location: 1 cun above the midpoint of the eyebrow.
Indications: Facial paralysis, headache, trigeminal neuralgia.
Point: Head-Linqi (G15)
Location: 0.5 cun above anterior hairline, between Pt. Shengting and Pt. Touwei.
Indications: Vertigo, stuff nose, nebula, apoplexy, coma, malaria, epilepsy, acute or chronic conjunctivitis, etc.
Point: Muchuang (G16)
Location: 1 cun posterior to Pt. Head-Linqi.
Indications: Headache, dizziness, swelling of head and face, conjunctivitis, toothache, apoplexy, etc.
Point: Zhengying (G17)
Location: 1 cun posterior to Pt. Muchuang.
Indications: Rigidity of the neck, dizziness, and vertigo, toothache, vomitting, etc.
Point: Chengling (G18)
Location: 1 cun posterior to Pt. Zhengying.
Indications: Headache, common cold, bronchitis, eye diseases, epistaxis, stuffy nose.
Point: Naokong (G19)
Location: On the lateral side of the protuberantia occipitalis externa, Pt. Fengchi is just below it.
Indications: Headache, common cold, asthma, epilepsy, mental diseases, palpitation, tinnitus, etc.
Point: Fengchi (G20)
Location: At the level of Pt. Fengfu, in the depression between m. trapezius and m. sternocleidomastoideus.
Indications: Common cold, headache, vertigo, stiffness and pain of the neck, eye diseases, rhinitis, tinnitus, deafness, hypertension, apoplexy, disorders of brain, etc.
Point: Jianjing (G21)
Location: At the middle point between Pt. Dazhui and the acromion, at the highest point of the shoulder.
Indications: Diseases of the shoulder and the back, mastitis.
Point: Yuanye (G22)
Location: On the mid-axillary line, at the level of the fourth intercostal space.
Indications: Pleuritis, intercostal neuralgia, lymphadenitis of the axilla, pain in the shoulder and back.
Point: Zhejin (G23)
Location: 1 cun below Pt. Yuanye.
Indications: Pleuritis, asthma, vomiting, regurgitation of acid.
Point: Riyue (G24)
Location: Directly under the papilla mammae in the 7th intercostal space.
Indications: Gastralgia, hepatitis, cholecystitis, disorders of the shoulder.
Point: Jingmen (G25)
Location: On the lower border of the free end of the 12th rib.
Indications: Nephritis, painful hernia, intercostal neuralgia, lumbago, pain of the leg, etc.
Point: Daimai (G26)
Location: Directly below Pt. Zhangmen, level with umbilicus.
Indications: Irregular menses, leucorrhea, hernia, endometritis, cystitis, pain of back and loin.
Point: Wushu (G27)
Location: At the lateral abdomen in front of the spina aliac anterior superior, level with Pt. Guanyuan.
Indications: Endometritis, leucorrhea, painful hernia, ochitis, lumbago, etc.
Point: Weidao (G28)
Location: Anterior and inferior to the spinae iliac anterior superior, 0.5 cun anterior and inferior to Pt. Wushu.
Indications: Adnexitis, endometritis, prolapse of the uterus, painful intestinal hernia, habitual constipation, etc.
Point: Femur-Juliao (G29)
Location: In the center of the line between the spina iliac anterior superior and the trochanter major of the femur.

CLASSICAL APs PERTAINING TO THE GALL BLADDER MERIDIAN (G) *(cont'd)*

Indications: Gastralgia, lower abdominal pain, orchitis, endometritis, cystitis, disorders in the hip joint and its surrounding soft tissue, lumbago and pain in the leg.

Point: Huantiao (G30)
Location: At the junction of lateral 1/3 and medial 2/3 of the line between the top point of the trochanter major of the femur and chiatus sacralis.
Indications: Sciatic neuralgia, lumbago and pain of the leg, numbness of the lower extremeties and paralysis.

Point: Fengshi (G31)
Location: With the patient standing erect, hands close to thigh, the point is at the tip of the middle finger.
Indications: Paralysis of lower extremeties, lumbago, pain in the leg, inflammation of the cutaneus nerve of lateral aspect of the thigh.

Point: Femur-Zhongdu (G32)
Location: 2 cun under Pt. Fengshi
Indications: Hongkong foot, paralysis and numbness of the lower extremities, sciatic neuralgia, etc.

Point: Xiyangguan (G33)
Location: In the depression superior to the condylus lateralis of the femur, let the patient flex the knee when locating, 3 cun above Pt. Yanglingquan.
Indications: Pain in the knee joint, numbness of the lower extremities and paralysis.

Point: Yanlingquan (G34)
Location: In the depression anterior and inferior to the caput fibulae, flex the knee during location.
Indications: Pain in the knee joint, sciatic neuralgia, hemiplegia, pain in the lower chest, cholecystitis, numbness of lower extremities.

Point: Yangjiao (G35)
Location: 7 cun above the malleolus lateralis on the anterior border of the fibula.
Indications: Fullness and pain in the lower chest, pain in the knee, weakness and atrophy of the foot, frenzy, edema of the face, etc.

Point: Waiqiu (G36)
Location: 7 cun above the malleolus lateralis on the anterior border of the fibula.
Indications: Headache, hepatitis, paralysis of the lower extremities.

Point: Guangming (G37)
Location: 5 cun above the top of the malleolus lateralis on the anterior border of the fibula.
Indications: Night blindness, optic atrophy, migraine, pain in the lateral aspect of the leg.

Point: Yangfu (G38)
Location: 4 cun above the top of the malleolus lateralis, on the anterior border of the fibula.
Indications: Migraine, cervical lymphadenitis, hemiplegia, numbness of the lower extremities, arthritis of the knee joint, etc.

Point: Xuanzhong (G39)
Location: 3 cun above the top of the malleolus lateralis, on the anterior border of the fibula.
Indications: Pain in the knee, the ankle and the lower chest, stiff neck, mehiplegia, sciatic neuralgia.

Point: Qiuxu (G40)
Location: Anterior and inferior to the malleolus lateralis, in the depression on the tendon of m. extensor digitorum longus.
Indications: Pain in the lower chest, cholecystitis, cervical lymphadenitis, sciatic neuralgia, disorders of ankle and its surrounding soft tissue, etc.

Point: Foot-Linqi (G41)
Location: In the depression anterior to the junction of the ossa metatarsalia IV and V.
Indications: Migraine, mastitis, pain in the lower chest, disorders of the lateral aspect of the lower extremities and dorsum of the foot.

Point: Diwuhui (G42)
Location: Between the ossa metatarsalia IV and V, 0.5 cun anterior to Pt. Foot-Linqi.
Indications: Tinnitus, mastitis, lumbago, swelling and pain in the dorsum of the foot.

Point: Xiaxi (G43)
Location: On the crevice between the fourth and fifth toes, proximal to the margin of the web.
Indications: Migraine, hypertension, tinnitus, intercostal neuralgia, etc.

Point: Foot-Qiaoyin (G44)
Location: On the lateral side of the tip of the fourth toe, 0.1 cun proximal to the corner of the nail.
Indications: Headache, hypertension, conjunctivitis, intercostal neuralgia asthma, pleuritis, etc.

CLASSICAL APs PERTAINING TO THE LIVER MERIDIAN (LIV)

Point: Dadun (Liv1)
Location: On the lateral aspect of the dorsum of the big toe, 0.1 cun proximal to the corner of the nail.
Indications: Prolapse of the uterus, painful hernia, menorrhagia, enuresis.

Point: Xingjian (Liv2)
Location: On the web between the first and second toes.
Indications: Irregular menstruation, amenorrhea, headache, insomnia, mental diseases, epilepsy, convulsion in children.

Point: Taichong (Liv3)
Location: On the dorsum of the foot between the ossa metatasale I and II, in the depression posterior to the art, metatarso-phalangee.
Indications: Headache, dizziness or vertigo, hypertension, irregular menstruation, menorrhagia, mastitis.

Point: Zhongfeng (Liv4)
Location: Anterior to the malleolus medialis, between the tendon of m. tibialis anterior and m. extensor hallucis longus.
Indications: Nocturnal emmision, dysuria, hernia, lumbago.

Point: Ligou (Liv5)
Location: 5 cun above the malleolus medialis, on the medial surface of the tibia near the medial border.
Indications: Irregular menstruation, dysurea, pain in the leg.

Point: Tibia- Zhongdu (Liv6)
Location: 7 cun above the malleolus medialis.
Indications: Irregular menstruation, menorrhagia, painful hernia, pain of the lower abdomen, pain in the joint of lower extremities.

Point: Xiguan (Liv7)
Location: In the posterior and inferior aspect of the condylus medialis of the tibia, 1 cun posterior to Pt. Yinlingquan.
Indications: Pain in the medial apsect of the knee, sore throat, pain or numbness caused by cold, damp, etc.

Point: Ququan (Liv8)
Location: In the depression at the medial end of the transverse crease of the art. genus with knee flexed.
Indications: Prolapse of the uterus, pruritus vulvae, dysuria, nocturnal emission, pain in the knee and in the medial aspect of the thigh.

Point: Yinbao (Liv9)
Location: 4 cun above the epicondylus mediallis, of the femur, between the m. gracilis and the m. satorius.
Indications: Irregular menstruation, incontinence of urine, retention of urine, lumbago, etc.

Point: Femur-Wuli (Liv10)
Location: 3 cun below Pt. Qichong lateral to the m. adductor longus.
Indications: Retention of urine, drowsiness, enuresis, eczema of the scrotum, pain in the medial aspect of the thigh, etc.

Point: Yinlian (Liv11)
Location: 2 cun below Pt. Qichong.
Indications: Irregular menstruation, pain in the lower extremities, painful hernia.

Point: Jimai (Liv12)
Location: Lateral and inferior to the tuberculum pubicum 2.5 cun lateral to the Ren Mai.
Indications: Prolapse of the uterus, painful hernia, hydrocele of the testis, pain in the penis, etc.

Point: Zhangmen (Liv13)
Location: At the lower border of the free end of the eleventh rib.
Indications: Abdominal distension and borborygmus, vomiting, diarrhea, jaundice, pain in the lower chest and the back, etc.

Point: Qimen (Liv14)
Location: Directly below the nipple in the sixth intercostal space; let the patient lie on the back while taking the point.
Indications: Intercostal neurlagia, hepatitis, enlargement of the liver, cholecystitis, pleurisy, gastric neurosis, etc.

CLASSICAL APs PERTAINING TO THE GOVERNING VESSEL (GV)

Point: Changqiang (GV1)
Location: Midway between the tip of the coccyx and the anus.
Indications: Prolapse of anus, passing blood in one's stool (blood stool), lumbago.

Point: Yaoshu (GV2)
Location: Below the fourth vertebra sacralis in the hiatus sacralis.
Indications: Irregular menstruation, pain in the lumbo-sacral region, epilepsy, cysts in the uterus or its appendages, cysts in the ovary, may be used for acupuncture anesthesia.

Point: Yaoyangguan (GV3)
Location: Between the processus spinosus of the fourth and fifth vertebrae lumbales.
Indications: Lumbago, paralysis of the lower extremities, irregular menstruation, nocturnal emission, impotence, etc.

Point: Mingmen (GV4)
Location: Between the processus spinosus of the second and third vertebrae lumbales.
Indications: Nocturnal emission, impotence, menorrhalgia, irregular menstruation, leucorrhea, chronic diarrhea, lumbago, frequently used for acupuncture anestheria in gynecological surgery.

Point: Xuanshu (GV5)
Location: Below the processus spinosus of the first vertebra lumbales.
Indications: Dysentery, abdominal pain, diarrhea, prolapse of anus, rigidity and pain in the lumbar vertebrae.

Point: Jizhong (GV6)
Location: Below the processus spinosus of the eleventh verterbra thoracica.
Indications: Hepatitis, epilepsy lumbago, paralysis of the lower extremities.

Point: Zhongshu (GV7)
Location: Below the processus spinosus of the tenth vertebra thoracica.
Indications: Gastralgia, cholecystitis, failing eye sight, lumbago, etc.

Point: Jinsuo (GV8)
Location: Below the processus spinosus of the ninth vertebra thoracica.
Indications: Hepatitis, cholecystitis, pleurisy, intercostal neuralgia, etc.

Point: Zhiyang (GV9)
Location: Between the processus spinosus of the seventh and eighth vertebrae thoracicae.
Indications: Jaundice, cough, asthma, malaria, pain in the vertebra thoracica, fullness of the lower chest, paralysis or atrophy of muscle, etc.

Point: Lingtai (GV10)
Location: Between the processus spinosus of the sixth and seventh vertebrae thoracicae.
Indications: Cough, backpain, painful swelling due to furuncle.

Point: Shendao (GV11)
Location: Below the processus spinosus of the fifth vertebra thoracicae.
Indications: Febrile diseases, heart disease, malaria, epilepsy, intercostal neuralgia.

Point: Shenzhu (GV12)
Location: Between the processus spinosus of the third and fourth vertebrae thoracicae.
Indications: Cough, dyspnea, epilepsy, pain in the back and the neck.

Point: Taodao (GV13)
Location: Between the processus spinosus of the first and second vertebrae thoracicae.
Indications: Fever, malaria, mental disease, epilepsy, etc.

Point: Dazhui (GV14)
Location: Between the processus spinosus of the seventh vertebrae cervicales and the first vertebra thoracicae.
Indications: Fever, malaria, common cold, cough, asthma, urticaria, stiff back, stiff neck, also used for preventing diseases and promoting health protection.

Point: Yamen (GV15)
Location: 0.5 cun above the mid-point of posterior hairline, directly below the protuberantia occipitalis externa.
Indications: Mental disease, epilepsy, sequelae apoplexy, cerebral concussion, chronic sore throat, deafness, mutism.

Point: Fengfu (GV16)
Location: 1 cun above the mid-point of posterior hairline, in the depression below the protuberantia occipitalis externa.
Indications: Mental disease, sequelae of apoplexy.

Point: Naohu (GV17)

Continued

CLASSICAL APs PERTAINING TO THE GOVERNING VESSEL (GV) *(cont'd)*

Location: 1.5 cun directly above Pt. Fengfu, on the upper margin of the protuberantia occipitalis externa.
Indications: Headache, stiff neck, insomnia, epilepsy.
Point: Qiangjian (GV18)
Location: 1.5 cun above Pt. Naohu.
Indications: Same as Pt. Naohu
Point: Houding (GV19)
Location: 1.5 cun above Pt. Baihui.
Indications: Head diseases.
Point: Baihui (GV20)
Location: At the junction between the line connecting the apexes of both ears and the top point of the sutura sagittalis.
Indications: Faint, headache, dizziness or vertigo, mental disease, prolapse of uterus, prolapse of rectum, usually used for acupuncture anaesthesia in brain surgery.
Point: Qianding (GV21)
Location: On the midline of the vertex, 1.5 cun anterior to Pt. Baihui.
Indications: Diseases of head region.
Point: Xinhui (GV22)
Location: 3 cun anterior to Pt. Baihui.
Indications: Headache, dizziness or vertigo, rhinitis, nasal polypi, convulsion in children.
Point: Shangxing (GV23)
Location: On the midline of the head, 1 cun above anterior hairline.
Indications: Headache, mental disease, disorders of the nasal cavity.
Point: Shenting (GV24)
Location: On the midline of the head, 0.5 cun above anterior hairline.
Indications: Headache, dizziness or vertigo, rhinitis, nasal polypi, mental disease.
Point: Suliao (GV25)
Location: In the middle of the apex nasi.
Indications: Shock, hypertension, bradycardia, brandy nose, epistaxis, rhinitis, etc.
Point: Renzhong (GV26)
Location: At the junction of the upper 1/3 and lower 2/3 of the nasal labial groove.
Indications: Shock, coma, mental disease, sunstroke, asphyxia, weakness of breath.
Point: Duiduan (GV27)
Location: On the median tubercle of the upper lip.
Indications: Vomiting, stuffiness of the nose, nasal polypi, epilepsy, stomatitis.
Point: Yinjiao (GV28)
Location: Take up the upper lip, at the labial end of the frenelum labii superior, slightly above the cleft between the incisors.
Indications: Acute sprain of lumbar region, nasal polypi, toothache, gum bleeding, mental disease.

CLASSICAL APs PERTAINING TO THE CONCEPTION VESSEL (CV)

Point: Huiyin (CV1)
Location: In the centre of the perineum.
Indications: Nocturnal emission, prostatitis, asphyxia, respiratory failure.
Point: Qugu (CV2)
Location: 5 cun below the umbelicus, on the superior border of the symphysis pubis, locate this point in supination.
Indications: Nocturnal emission, impotence, leucorrhea, anuria, hernia.
Point: Zhongji (CV3)
Location: 1 cun above Pt. Qugu, 4 cun below the umbelicus, locate this point when lies on back.
Indications: Nocturnal emission, uroclepsia, dysuria, frequent urination, gonorrhea, pain in lower abdomen, irregular menstruation, leucorrhea.
Point: Guanyuan (CV4)
Location: 2 cun above Pt. Qufu, 3 cun below the umbilicus, locate when lies on back.
Indications: Irregular menstruation, impotence, uroclepsia, abdominal pain, dysentery amenorrhea, metrorrhagia, leucorrhea, prolapse of the uterus, intestinal ascarisis, etc.
Point: Shimen (CV5)
Location: On the front midline 2 cun below the umbilicus.
Indications: Metrorrhagia, amenorrhea, diarrhea, edema, hypertension.
Point: Qihai (CV6)

CLASSICAL APs PERTAINING TO THE CONCEPTION VESSEL (CV) (cont'd)

Location: 1.5 cun below umbilicus, locate this point when lies on back.
Indications: Nocturnal emission, impotence, irregular menstruation, menorrhalgia, distension, diarrhea, retention of urine, frequent urination.
Point: Abdomen-Yinjiao (CV7)
Location: 1 cun below the umbilicus.
Indications: Metorrhagia, leucorrhea, irregular menstruation, pruritus vulvae, peri-umblical pain, hernia, post partum bleeding.
Point: Shenjue (CV8)
Location: In the centre of the umbilicus, locate this point when lies on back.
Indications: Acute enteritis (of cold nature), chronic enteritis, chronic dysentery, intestinal tuberculosis, edema, etc.
Point: Shuifen (CV9)
Location: 1 cun above the umbilicus.
Indications: Borborygmus, diarrhea, abdominal pain, edema, dysuria, swelling of the face.
Point: Xiawan (CV10)
Location: 2 cun above the umbilicus, locate this point when lies on back.
Indications: Gastralgia, vomiting, abdominal distension, dysentery.
Point: Jianli (CV11)
Location: 1 cun below Pt. Zhongwan
Indications: Gastralgia, vomiting, anorexia, abdominal distension and edema.
Point: Zhongwan (CV12)
Location: 4 cun above the umbilicus, locate this point in supine position.
Indications: Gastralgia, vomiting, hiccup, abdominal distension, diarrhea, gastritis gastric ulcer, gastric ptosis, acute intestinal obstruction, etc.
Point: Shangwan (CV13)
Location: 5 cun above the umbilicus, 1 cun above Pt. Zhongwan.
Indications: Gastritis, gastric dilatation, gastric spasm, cardiac spasm.
Point: Juque (CV14)
Location: 6 cun above the umbilicus on the frontal midline.
Indications: Gastralgia, hiccup, palpitation, mental disorder, epilepsy.
Point: Juiwei (CV15)
Location: 7 cun above the umbilicus, 1 cun below the processus xiphoideus of the sternum.
Indications: Pain in the cardiac region, hiccup, mania, epilepsy.
Point: Zhongting (CV16)
Location: On the midsternal line at the level of the fifth intercostal space.
Indications: Fullness of lower chest, hiccup, regurgitation of food, infantile milk regurgitation.
Point: Shanzhong (CV17)
Location: Midway between the two nipples, locate this point with patient in supine position.
Indications: Chest pain, fullness of chest, diffience of lactation, intercostal neuralgia, angina pectoris, asthma, etc.
Point: Yutang (CV18)
Location: On the midsternal line at the level of the third intercostal space.
Indications: Cough, asthma, pain in the chest, paralysis of larynx, obstruction in the larynx, excessive expectoration.
Point: Chest-Zigong (CV19)
Location: On the midsternal line at the level of the second intercostal space.
Indications: Cough, asthma, pain in the chest, paralysis of pharynx, obstruction in the larynx.
Point: Huagai (CV20)
Location: On the midsternal line at the level of the first intercostal space.
Indications: Asthma, cough, fullness and pain in the lower chest.
Point: Xuanji (CV21)
Location: On the midsternal line at the level of superior margin of the first rib.
Indications: Cough, asthma, pain in the chest.
Point Tiantu (CV22)
Location: In the depression on the suprasternal fossa.
Indications: Bronchitis, asthma, pharyngitis, goiter, spasm of the diaphragm, diseases of the esophagus, neurological vomiting, etc.
Point: Lianquan (CV23)
Location: Above the prominentia laryngea, in the depression of the superior margin of the os hyoideum.
Indications: Bronchitis, pharyngitis, muteness, asthma, excessive salivation.
Point: Chengjiang (CV24)
Location: On the mid-mandibular line, in the depression below the mental labial groove.
Indications: Trigeminal neuralgia, facial paralysis.

5.8.1.2 Basic Principles of Modern Acupuncture

5.8.1.2.1 SELECTION OF THE MOST SUITABLE APs

All points chosen for therapeutic stimulation should display an increased sensitivity to physical pressure (practical, but subjective criterion) and a high degree of rectification/increased impedance phenomenon (objective scientific criterion) (see Sections 3.3 and 4.4.2). In practice, a precise location of the required AP should be established by a moderately strong palpation of the entire skin region in which the AP in request is supposed to be found according to the classical acupuncture maps; only significantly tender skin spots should be considered for acupuncture stimulation. Tenderness of APs related to the diseased organ is caused by the presence of high-frequency action potentials (originating from the organ) in the sensory nerves innervating these skin areas. A high degree of rectification/increased impedance phenomenon will therefore also be present there (see Section 3.3), which can be confirmed with an OED device; only APs regarded as "Acute" or "Subacute" should be stimulated. During the successful course of treatment, the tenderness of many APs will gradually subside, as will the high degree rectification/increased impedance phenomenon. This means that the total number of needles used can be diminished at the end of the course of treatment.

When treating a particular internal organ/body part, tender local classical APs and other local TPs/PPs should be first of all considered for a therapeutic stimulation. Distal APs located along the radiation of pain—usually on meridians passing through the appropriate anatomical region—should also be used as long they display a significantly increased tenderness (see earlier explanation).

When treating any problem of spinal origin, such as upper back pain with or without brachialgia, lower back pain with or without sciatica, intercostal neuralgia, cervicogenic headache, etc., tender classical APs as well as other TPs/PPs situated on the respective dermatomes (Figure 5.35) should be stimulated.

Skin spots considered for stimulation should be at least 2–3 cm apart from each other, with the exception of the auricular OPAs, which are located very densely on the surface of the ear auricle.

The more APs are simultaneously stimulated, the better the chance to block the respective afferent nervous signals carried by thin "nociceptive" fibers. However, if the stimulation is too diffused, then the therapeutic effect might be diminished due to the adaptation of the nervous system. Therefore, in practice, only the most tender 10 to 30 relevant skin areas (depending on the problem) should be stimulated at the same time.

Marking the locations of particular APs for future use (such as with a ballpoint pen) does not work in practice, because even a small change of the body's position will affect the location of these skin spots. Therefore, the precise locations of required APs and other PPs have to be established every time immediately before stimulation, according to the previously mentioned criteria.

5.8.1.2.2 ACUPUNCTURE NEEDLES AND PROCEDURES

The needles should be of adequate durability and resilience and resistant to high temperatures and chemical substances used for sterilization. Historically, sharp pieces of stone, bone, or bamboo and even thorns were initially used for therapeutic purposes; golden and silver needles were once used and were believed to have special stimulating/sedative features. High-quality alloy

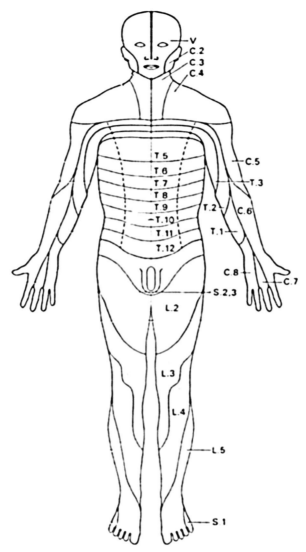

FIGURE 5.35 A Dermatomes (anterior aspect). *From the Oxford Handbook of Clinical Medicine, 8th ed., by Longmore, Wilkinson, Davidson, Foulkes, & Mafi (2010); reproduced by permission of Oxford University Press.*

needles are in contemporary use, typically with a relatively thicker hilt and a very thin spike to be inserted into the skin (Figure 5.36). Acupuncture needles are usually not extremely sharp, to prevent any potential tissue damage. The standard needle length of 3–4 cm is useful for thin skin (such as that on the face, ear auricles, or fingers), 6.5–10 cm is usually reserved for bigger muscle areas (such as the lower back), and 5 cm is generally used for all other regions of the body. Acupuncture needles are supposed to stimulate skin nerves; therefore, only the needle tip needs to reach the subcutaneous tissues. Small round "microneedles/press-needles," with very short spikes, can be placed on the ear auricle and left there for seven to ten days after fixing them in place with a piece of self-adhesive tape. Because acupuncture needles are solid pins, and not thin pipes like syringe needles, the risk of an infection transfer is relatively low; it is known that germs can survive the skin's self-defense mechanisms inside a syringe needle. Nevertheless, the use of disposable sterile acupuncture needles for single use is most highly recommended; otherwise, strict general sterilization rules have to be followed. Before insertion and after removal of the needle, the skin must always be cleaned with an appropriate disinfecting fluid.

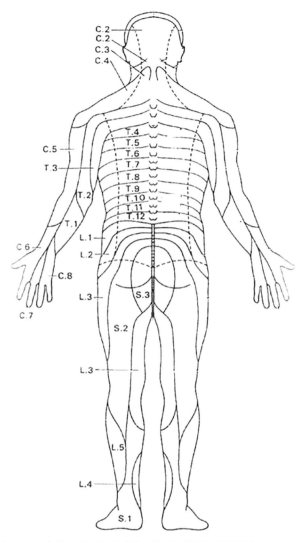

FIGURE 5.35 B Dermatomes (posterior aspect). *From the Oxford Handbook of Clinical Medicine, 8th ed., by Longmore, Wilkinson, Davidson, Foulkes, & Mafi (2010); reproduced by permission of Oxford University Press.*

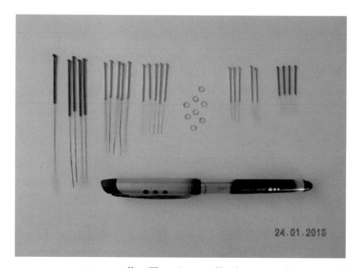

FIGURE 5.36 Typical contemporary acupuncture needles. The microneedles/press needles are visible in the middle of the illustration.

The needle inserted into the skin should be gently manipulated until the so-called jump sign (De qi) occurs; this is a sudden shake-up of the treated part of the body accompanied by a mild electric shock-like sensation felt by the patient (sometimes it can be also felt as heaviness, tingling, numbness, or another feeling). The jump sign, presumably caused by directly touching the skin nervous receptors, is a practical indication of the correct insertion of the acupuncture needle into an AP, whereas the simultaneous sudden drop in electrical potential measured on the needle (see Section 2.5.2) can serve as objective evidence. Obtaining the jump sign is considered to be essential for the therapeutic efficacy of acupuncture. If needles are used for a single treatment, they should be rotated every few minutes to prevent adaptation of the nervous receptors; the typical duration of the acupuncture session is 20 to 30 minutes. Sometimes the needle inserted into a muscle spasm can hurt; once the patient relaxes the muscle can pull on the needle. In such a case, a small manual adjustment of the needle can make it much more comfortable. Nevertheless, in order to avoid any problems of this kind, all patients undergoing an acupuncture treatment should not move once the needles are inserted into the skin. It is observed that many patients treated for painful conditions complain more about "painful needles" once their conditions get better; at the beginning, they do not feel the needles due to too much of their own pain. It also quite often occurs that patients faint at the first or second acupuncture session, especially when treated in the sitting position; this has nothing to do with the treatment as such, it is just a common psychological reaction to needles.

Traditional acupuncture rules stipulate that needles should be inserted first in the distal parts of the body and only then in the proximal areas. And vice-versa: The needle removal should start with needles inserted in proximal areas and finish with distal ones. This rule has some physiological basis in the specific structure of the sensory nervous system. Slight bleeding, which sometimes follows the removal of needles, can be easily stopped by applying moderate pressure to the wound with a piece of cotton wool soaked in disinfecting fluid.

5.8.1.2.3 COMBINED ACUPUNCTURE TREATMENTS

In order to obtain a maximal amplitude and intensity of the therapeutic stimulus as well as prevent an adaptation of the nervous system, it is good to combine needles with other kinds of stimulation, such as electrotherapy or thermotherapy.

Electroacupuncture (Figure 5.37) is perhaps the most powerful therapeutic method of physical medicine. In general, the needles inserted into the skin can be directly stimulated with the same currents as used for TENS (see Section 5.6.1), but intensity has to be two to three times weaker; it should be adjusted individually between the sensitivity and the pain thresholds. This is because the electrical resistance of the needle inserted into the skin is much lower than that measured on the surface of undamaged skin. A fully balanced bipolarity of the impulse is especially important, because electrocoagulation can occur much more easily on inserted needles than on the dry surface of undamaged skin. Only the most important APs (usually the local ones) should be stimulated with electroacupuncture.

Traditionally, needles can be combined with moxibution (see Section 5.2.1). However, even better results can be obtained with a cryotherapeutic stimulation of the needles (−160 or −70 degrees Celsius), because the amplitude of a cryotherapeutic stimulus is much higher; the equipment described in Section 5.2.2 can be used, mainly for the most important APs. Because metal is a very good thermal conductor, the duration of such a cryopuncture is usually shorter. For practical

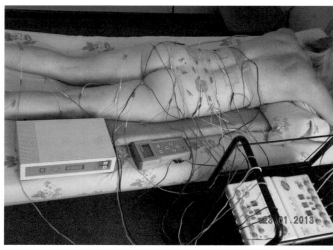

FIGURE 5.37 Complex physical therapy: classical acupuncture, electroacupuncture, TENS, and pulsed magnetic field therapy used simultaneously for lower back pain. Cryopuncture to follow after removal of cables and TENS electrodes.

purposes, it is best to perform cryotherapeutic stimulation of acupuncture needles at the end of a combined treatment: for example, after completion of the electrical stimulation and removal of the cables.

Acupuncture procedures, including combined treatments, can be performed simultaneously with other methods of physical medicine such as TENS or magnetic field therapy. The typical duration of combined acupuncture treatments is up to 45–50 minutes, including precise manual localization of particular APs and insertion of the needles or placing of the TENS electrodes, connecting the chosen needles to the electrostimulators and adjusting the current intensity, 20–30 minutes of therapeutic stimulation, and removal of cables and needles or TENS electrodes. If a cryotherapeutic stimulation of needles is also to be used after electrostimulation, a few additional minutes should be added. The typical course of combined acupuncture therapy includes at least 20 sessions in order to ensure a long-lasting improvement; if the treatment is finished too early, the problem may come back within two to three months. However, when dealing with something such as an advanced neuropathy, the treatment may take 30– 40 sessions or even longer; it takes time to regenerate the nerves, with the first signals of clinical improvement starting only after the basic course of 20 sessions. During the acute phase, the treatments can be performed on a daily basis; chronic cases can be treated three times per week.

5.8.1.2.4 GENERAL CONTRAINDICATIONS FOR ACUPUNCTURE

Acupuncture should not be performed directly on malignant tumours. However, it can be successfully used for cancer pains and will not increase the probability of metastases, because malignant cells are already present in microcirculation. Acupuncture needles should not be inserted in moles, warts, and pathologically changed skin. Acupuncture of the lower back, abdomen, and legs should be avoided in pregnancy due to the risk of inducing premature delivery. Anticoagulative therapy and well-controlled hemophilia are only mild contraindications for acupuncture, because thin acupuncture needles are very seldom a cause of bleeding. The use of disposable sterile needles or preservation of strict general sterilization rules (when using the same needles for the whole course of treatments) can prevent the risk of HIV or other infections.

5.8.1.3 Chosen Examples of Acupuncture Therapy

5.8.1.3.1 PAIN SYNDROMES OF SPINAL ORIGIN: UNDERSTANDING PATHOPHYSIOLOGY

Pain syndromes of spinal origin (PSSO) typically include back pains (lower/dorsal/upper) with or without neuralgias: sciatica, L2 neuralgia, L1 neuralgia, T11–12 neuralgia, intercostal neuralgia, brachialgia, C3 neuralgia, or occipital neuralgia. However, it seems that 85–90% of chronic headaches could also be of spinal origin. *One will never die because of spinal pain, but one will also not want to live because of spinal pain*; it can be an excruciating and disabling sensation that can ruin one's life. Yet there are practically no adults who have never experienced any back pain whatsoever. Also, from an economic point of view, back problems are among the most important health issues in the world; most sick leaves are issued and most disability grants are given due to back problems.

Nonethless, there is a lot of controversy surrounding the pathophysiology of persistent back pains; different explanations are given by neurologists, orthopedists, neurosurgeons, rheumatologists, chiropractors, physiotherapists, and even psychologists. In fact, not everybody with evident changes on spinal X-rays/scans suffers pain, and vice versa. Many, especially young people, with perfect X-rays/scans cannot even move because of back pain; this cannot be directly explained by orthopedists and neurosurgeons. In the latter case, the OED appears to be very useful in verification of the presence of strong afferent nervous signals originating from the respective parts of the spine. Doctors/therapists who do not have access to OED technology can still confirm spinal problems by properly carried out physical examinations of the spinal column; moderately strong palpation along the spine will reveal very tender areas at the respective spinal levels.

PSSO are generally caused by a persistent physical pressure applied to the nerves entering the spinal cord (nerve roots) and sometimes the cord itself by prolapsed intervertebral discs and/or degeneratively deformed vertebral elements; local tumours, fractures, epidural abscesses, spinal TB, or other infectious/inflammatory spinal problems are statistically very rare. Therefore, some specialists rightfully describe these kinds of pains as "mechanical back pains." Pinched nerves become inflamed; it is important, however, to remember that the pain automatically triggers a nervous reflex arch, which results in the strong spasm of the surrounding muscles, especially those innervated by the respective nerves. The typical vicious circle is thus created; the pain causes muscle spasms, and muscle spasms significantly increase the pressure applied to pinched nerves, which in turn causes the pain (see Section 5.1). Sometimes, in serious cases, reflexive muscle spasms include the internal organs in the neighborhood; severe lower back pain, for instance, can be a cause of spastic colon and/or neurogenic bladder. Prolonged overcontraction of muscles makes them painful on their own, leading to the so-called regional pain syndromes, with trigger points located along the related dermatomes. In addition, generalized muscle spasms cause the functional closing of small arteries supplying blood to the affected areas, including the pinched nerves and the intervertebral discs. With regard to the nerves, prolonged limitation of the blood supply can eventually lead to nerve degeneration (neuropathy), often with initial "pins and needles" and then numbness, especially at the distal parts of the respective nerves. With regard to the intervertebral discs, prolonged mechanical pressure and limitation of the blood supply will lead to flattening and finally to so-called dehydrated discs; narrow intervertebral spaces will also contribute to the increased pressure applied by the solid spinal elements to the nerve roots at, for example, the neural foramina level. Reflexive vasoconstriction of cerebral arteries, caused by pinched C1 and/or C2 nerve roots, will directly result in the so-called visual aura, which can develop to a full migraine (see Section 5.8.1.3.11). Both the prolonged muscle spasm and the limited local blood supply are perhaps major factors in restless leg syndrome. In general, it seems that the functional component of PSSO may be its leading pathophysiological factor; this could explain the phenomenon of persistent back pains in people who have perfect-looking X-rays/scans, but often also have a forgotten history of minor injuries or a course of "microinjuries" that triggered the "vicious circle." Mental stress can aggravate PSSO by strengthening muscle spasms, but it is not a primary cause!

Because persistent PSSO are directly connected with the functioning of the nervous system, physical medicine and in particular combined acupuncture therapy is very well positioned when it comes to the effective treatment of these problems. Blockage of strong nociceptive signals originating from pinched nerves at the local level of the spinal cord will break the vicious circle (see Sections 3.3 and 5.1) and result in three benefits: pain relief, muscle spasm relaxation, and improved local blood supply. Muscle spasm relaxation will prevent any further severe pressure being applied to the nerve roots and improved blood supply will slowly but surely regenerate damaged nerves, rehydrate intervertebral discs (at least to some extent) and heal certain infectious/inflammatory problems. Therefore, on completion of the successful course of treatment, one can expect a long-lasting clinical improvement even in serious spinal cases. Previous failed spinal surgeries make the task more difficult due to potential scar problems, but they are not a contraindication for intensive physical medicine treatment.

Simultaneous use of supportive therapies such as therapeutic exercises (stretching to relax muscle spasms) and proper medication (anti-inflammatories, pain killers, and antiepileptics in case of neuropathy) will accelerate the improvement, even if they did not help on their own. However, in the acute phase of severe spinal pain syndromes an immediate introduction of therapeutic exercises is not always possible; in such a case, exercises should be started at a later stage once improvement has been made through physical medicine and pharmacotherapy.

As mentioned previously, treatments can be done on a daily basis during the acute phase and three times per week in the case of chronic pain.

5.8.1.3.2 LOW BACK PAIN WITH OR WITHOUT SCIATICA

In our clinical material of hundreds of patients with severe low back pain, caused mainly by multilevel prolapsed discs (often with spinal stenosis) and/or significant degenerative vertebral changes, we have achieved an approximate 85% success rate with lasting complete or almost complete decreases in pain intensity on the analog pain scale, complete or almost complete drops in the amount of medication used, and significant improvements in mobility. Because the achieved improvement was usually long-lasting, many of our patients who were on disability pensions due to back problems were able to return to their jobs. Therefore, it seems that physical medicine should be the first choice when it comes to the management of severe back pain, especially because it can be used immediately in the acute phase—in contrast to therapeutic exercises, which, for obvious reasons, can only be started at a later stage. More invasive and risky treatments, such as surgical intervention, should be reserved for that statistically small group of patients who cannot be helped through physical medicine.

The selection of the most suitable APs should start with lying the patient down on his or her stomach and performing a moderately strong palpation along the lower spinal column; the presence of very sensitive/tender skin spots will precisely indicate the affected levels and therefore the

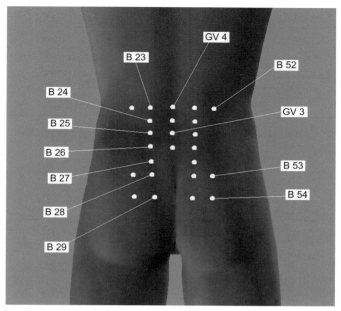

FIGURE 5.38   Recommended choice of classical APs and other PPs for the treatment of low back pain (PPs marked but not described).

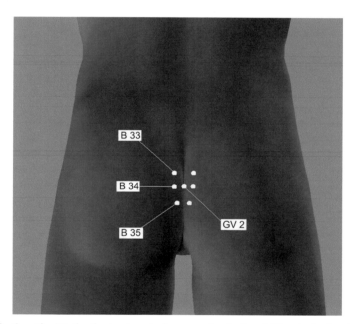

FIGURE 5.39   Specific APs for the treatment of coccygeal pain (in addition to the low back treatment).

respective dermatomes to work on. However, due to generalized muscle spasms and the creation of the so-called regional pain syndrome, some other dermatomes can be affected as well. Therefore, there are certain "universal" APs and other PPs that can be used practically in each case of low back pain, as long as they display increased sensitivity to palpation ("Acute" or "Subacute" readings on the OED device). These points may include (Figure 5.38): spots between spinous processes—L2 and L3 (GV 4), L3 and L 4, L4 and L5 (GV 3), and L5 and S1; and bilaterally—B 23 (B 24, B 25), B 26, B 27, B 28, B 29, B 52 and B 53 (both can be bilaterally stimulated with TENS), and B 54. In the case of severe coccygeal pain (Figure 5.39), GV 2 and bilaterally B 33, B 34, and B35 can be added. If the pain includes the genital and anal region (Figure 5.40), bilaterally B 32, B 33, and B 35; also, the use of GV 1 and CV 1 can be considered.

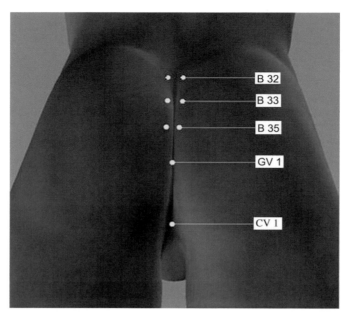

FIGURE 5.40 Specific APs for the treatment of the genital and anal regional PSSO.

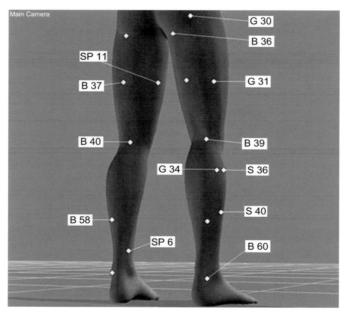

FIGURE 5.41 Specific APs for the treatment of sciatica (always in addition to the low back treatment).

If the pain radiates from the lower back to one or both legs (sciatica), the following APs can be added (Figure 5.41): B 36, B 37, B 39 or B 40, B 58, B 60, G 30, G 31, G 34, SP 6, and SP 11. In case of the pain originating predominantly from the L 4 level, the addition of S 40 and S 36 could be worthwhile. If only one leg is affected, the treatment can be applied unilaterally. If, however, the other leg is also affected (even slightly), the treatment should be applied symmetrically. Feet numbness caused by a neuropathy usually requires additional bilateral stimulation (Figure 5.42) of B 31, B 32, Liv 4, SP 3, G 40; K 1 and G 42 can be stimulated with TENS (same electric channel) and cryotherapy, for practical reasons. As mentioned earlier, the regeneration of severely damaged nerves can take much longer than the basic course of 20 therapeutic sessions.

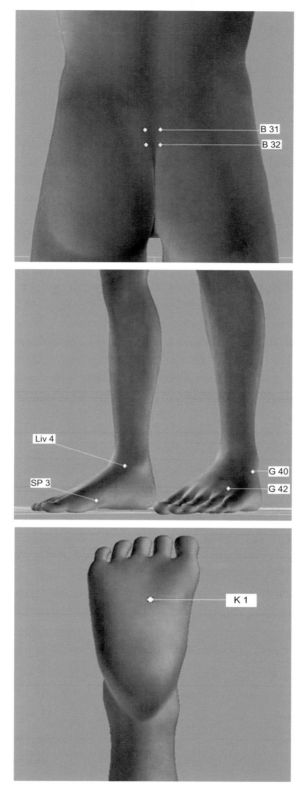

FIGURE 5.42 Specific APs for the treatment of foot numbness caused by neuropathy (in addition to the low back pain with sciatica treatment).

5.8.1.3.3 BACK PAIN IN PATIENTS WITH SPINAL FUSION

There are many patients with a history of spinal fusion who still suffer pain in the area which underwent surgery: This is called "failed spinal surgery syndrome." Even if the surgery was successful, within a few years the patients are likely to develop problems, typically above and below the fusion. In such a case, reflexive physical medicine and especially combined acupuncture therapy should be the first choice before consideration of any further surgery.

However, there might be technical problems concerning the postoperative scar or even posterior fixation; it does not make sense to insert needles into the scar because of lack of sensory innervation there. Therefore, if the scar is located along the governing vessel (GV), the respective spots located between spinous processes should be replaced with the so-called Huatuo points (usually very tender), located approximately 1.5 cm on both sides of the GV (Figure 5.43). The other respective APs can be still used as described in the previous section.

5.8.1.3.4 L2 NEURALGIA

L2 neuralgia is usually the part of the lower back regional pain syndrome but sometimes, especially in younger patients without complications, it can appear alone. It is usually caused by a pressure applied to the L2 nerve roots and results in persistent pain and increased sensation for L2 distribution (see Figure 5.35), just above the buttocks and sometimes the anterior aspects of the thighs. In our limited clinical material with this particular pain syndrome, we have achieved an approximate 85% success rate with a complete or almost complete decrease in pain intensity on the analog pain scale, improved mobility, and a complete or almost complete drop in the amount of medication used.

The treatment of a patient with a clear L2 neuralgia can be done in a sitting position, with insured access to his/her lower back. Moderately strong palpation along the spinal column will reveal increased sensitivity/tenderness of the L2–3 intervertebral space (GV 4). Palpation along the L2 dermatome will localize the respective PPs/TPs ("Acute" or "Subacute" readings on the OED device) for stimulation, including classical APs (Figure 5.44): B23 (always bilaterally; can be stimulated with TENS), B 52, G 29, S 31, SP 11. If only one side is affected, the treatment can be applied unilaterally. If the other side is also affected, however (even slightly), the treatment should be applied symmetrically.

5.8.1.3.5 L1 NEURALGIA WITH INGUINAL PAIN

Currently not all doctors/therapists are aware that a pressure applied to the L1 nerve roots may result in persistent groin pain along the L1 dermatome (see Figure 5.35); this can create diagnostic

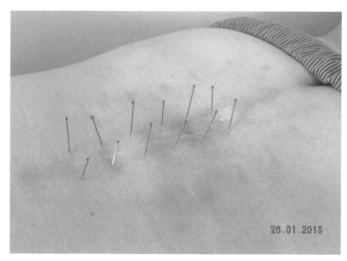

FIGURE 5.43   Acupuncture treatment of a patient with a post operative scar using the Huatuo points.

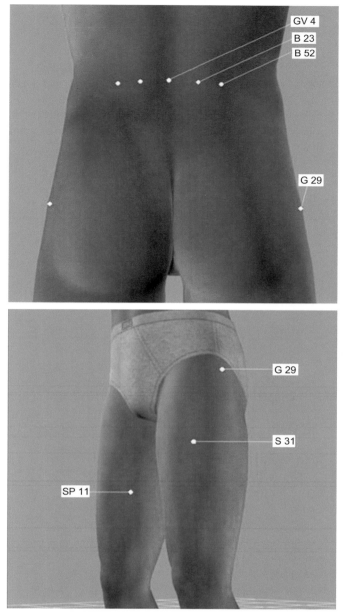

FIGURE 5.44 Recommended choice of APs for the treatment of L2 neuralgia.

problems once all other potential causes of inguinal pain are ruled out. In our clinical material of dozens of patients with this particular pain syndrome, we have achieved an approximate 80% success rate with a complete or almost complete decrease in pain intensity on the analog pain scale and a complete or almost complete drop in the amount of medication used.

Moderately strong palpation along the spinal column of the patient with L1 neuralgia will reveal increased sensitivity/tenderness of the L1–2 intervertebral space (GV 5). Palpation along the L1 dermatome will localize the respective PPs/TPs ("Acute" or "Subacute" readings on the OED device) for stimulation, including classical APs (Figure 5.45): B 22 (always bilaterally; can be stimulated with TENS), B 51, G 25, G 26, G27 or 28, S 30; because the patient undergoing treatment is usually lying on his/her stomach, it is practical to stimulate this point with TENS. If only one side is affected, the treatment can be applied unilaterally. If the other side is also affected, however (even slightly), the treatment should be applied symmetrically.

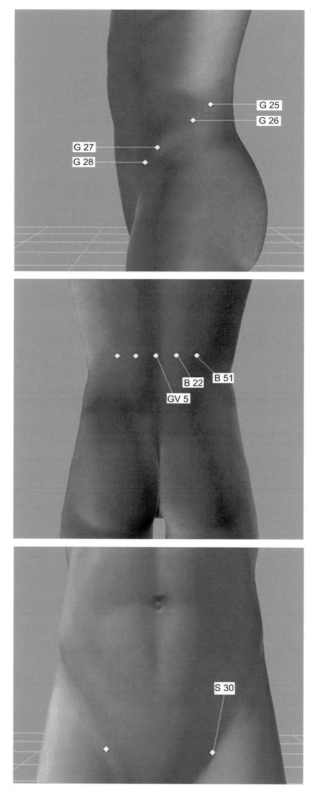

FIGURE 5.45 Recommended choice of APs for the treatment of L1 neuralgia.

5.8.1.3.6 T11-12 NEURALGIA WITH LATERAL ABDOMINAL PAIN

Currently not all doctors/therapists are aware that a pressure applied to the T11–12 nerve roots may result in the persistent lateral abdominal pain (see Figure 5.35); this can create diagnostic problems once all the other potential causes of pain in this region are ruled out. In our clinical material of dozens of patients with this particular pain syndrome, we have achieved an approximate 80% success rate with a complete or almost complete decrease in pain intensity on the analog pain scale and a complete or almost complete drop in the amount of medication used.

Moderately strong palpation along the spinal column of the patient with T11–12 neuralgia will reveal increased sensitivity/tenderness of the T11–12 (GV 6) or/and T12–L1 intervertebral spaces. Palpation along the T11 or/and T12 dermatomes (see Figure 5.35) will localize the respective PPs/TPs ("Acute" or "Subacute" readings on the OED device) for stimulation, including classical APs (Figure 5.46): B 20, B 21 (these points can be stimulated bilaterally with

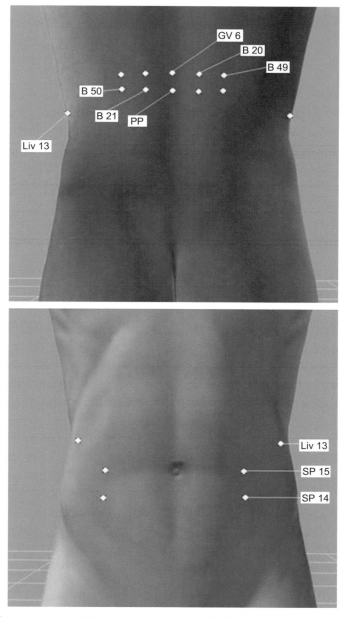

FIGURE 5.46 Recommended choice of classical APs and a PP for the treatment of T11–12 neuralgia.

TENS), B 49, B 50, Liv 13, SP 14, SP 15. If only one side is affected, the treatment can be applied unilaterally. If the other side is also affected, however (even slightly), the treatment should be applied symmetrically.

5.8.1.3.7 INTERCOSTAL NEURALGIA/POST-HERPETIC PAIN

Severe intercostal neuralgia can be a source of much stress to patients. On the diagnostic side, because of chest pain, it can be initially regarded as a serious cardiac, pulmonary, breast, or even hepatic or gastric problem. Once it becomes chronic, much too often it is labeled as so-called fibromyalgia (see Section 5.8.1.3.12), especially in younger patients with few changes on spinal X-rays. On the therapeutic side, the generally recommended medication (mainly antiepileptics and antidepressants) does not seem to be very effective, and nerve blockades carry a high risk of pneumothorax.

On most occasions, intercostal neuralgias are caused by pinched nerve roots at various levels of the thoracic spine; however, the most persistent, excruciating pains are those of the postherpetic neuralgia, with profoundly damaged intercostal nerves. Therefore, in the case of the postherpetic neuralgia even such radical methods as an ablation of the respective ganglion are tried, often with no success.

In our clinical material of hundreds of patients with severe intercostal neuralgia, we have achieved an approximate 90% success rate with a complete or almost complete decrease in pain intensity on the analog pain scale and a complete or almost complete drop in the amount of medication used. For dozens of cases of postherpetic neuralgias, our success rate equaled approximately to 80%, with a tolerable level of the remaining pain well controlled by low doses of medication. But it has to be emphasized that the average treatment time of postherpetic neuralgia was three to four times longer than that of a typical intercostal neuralgia; it takes much longer to regenerate seriously damaged nerves in the case of the postherpetic neuralgia.

For practical reasons, the treatment of intercostal neuralgias should be done in a sitting position. A moderately strong palpation along the thoracic spinal column will reveal the presence of very sensitive/tender spots between the spinous processes T4–T10; in this way, the affected levels and respective dermatomes will be precisely identified. In the case of postherpetic neuralgia, spinal column palpation is not always helpful, but the patient himself or herself and his or her skin discoloration will indicate the affected dermatome. Further palpation along the respective intercostal spaces will reveal the PPs/TPs ("Acute" or "Subacute" readings on the OED device), located approximately 3–5 cm apart from each other. Because the classical Chinese art of acupuncture does not provide any specific APs for these purposes, these PPs/TPs (Figure 5.47), including those between spinous processes, are to be stimulated in case of both intercostal and postherpetic neuralgias. Small acupuncture needles can be relatively safely used, in contrary to syringe needles which could easily cause a pneumothorax. However, in the case of female patients acupuncture of breasts should be avoided in order to minimize the risk of potentially inducing breast cancer; TENS and cryotherapy can be still used. It is good to remember that typical intercostal neuralgias, in contrary to postherpetic ones, very seldom affect only one dermatome; more often two or even three separate levels have to be treated simultaneously (regional pain syndrome).

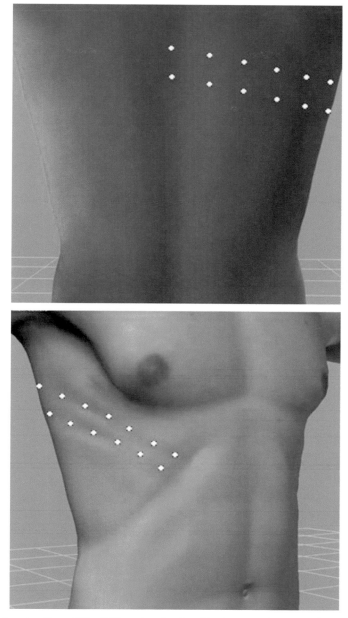

FIGURE 5.47   Exemplary location of PPs / TPs, to be stimulated in the case of intercostal neuralgia/post herpetic pain.

5.8.1.3.8 UPPER BACK PAIN WITH OR WITHOUT BRACHIALGIA

These days, upper back pain is becoming a more common condition, as it is an occupational hazard of people spending long hours with computers, usually with poor body positioning. In our clinical material of hundreds of patients with severe upper back pain, caused mainly by prolapsed discs and/or significant degenerative vertebral changes, we have achieved an approximate 85% success rate with a complete or almost complete decrease in pain intensity on the analog pain scale, a complete or almost complete drop in the amount of medication used, and a significant improvement in the neck and shoulders movement range.

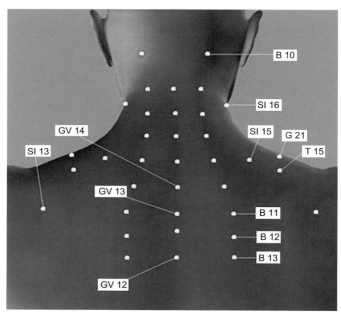

FIGURE 5.48 Recommended classical APs and other PPs / TPs to choose from when treating upper back pain (PPs and TPs marked but not described).

For practical reasons, the treatment of upper back pain should be done in a sitting position. Moderately strong palpation along the cervical spinal column will reveal a presence of very sensitive/tender spots between the spinous processes C3–T3; in this way, the affected levels and respective dermatomes will be precisely identified. However, due to the generalized muscle spasm and the creation of the so-called regional pain syndrome, some other dermatomes can be affected as well. Therefore, there are certain "universal" classical APs as well as other PPs/TPs that can be stimulated practically in each case of upper back pain, so long as they display increased sensitivity to palpation ("Acute" or "Subacute" readings on the OED device). These points (Figure 5.48) include, first of all, the previously mentioned tender spots between the cervical spinous processes as well as TPs located at the same levels, symmetrically on both sides of the neck, along the lines connecting APs B 10 and B 11. Among classical APs, the most useful are: SI 16 (can be stimulated bilaterally with TENS), SI 15, SI 13, B 10, B 11, B 12, B 13, G 21, T 15 (can be stimulated bilaterally with TENS), GV 12, GV 13, and GV 12.

If the pain radiates down the arm and the forearm (brachialgia) with "pins and needles" or even numbness in certain fingers, depending on the spinal level where the nerve roots are pinched, the use of the following APs should be considered (Figure 5.49): LI 17, LI 15, LI 14, LI 13, LI 10, LI 4, SI 9, T 5, and T 3. If only one side is affected, the treatment can be applied unilaterally. If the other side is also affected, however (even slightly), the treatment should be applied symmetrically.

The above described combined acupuncture therapy can also be used as an effective supportive treatment for Parkinsonism; early clinical improvement can usually be seen after only a few therapeutic sessions.

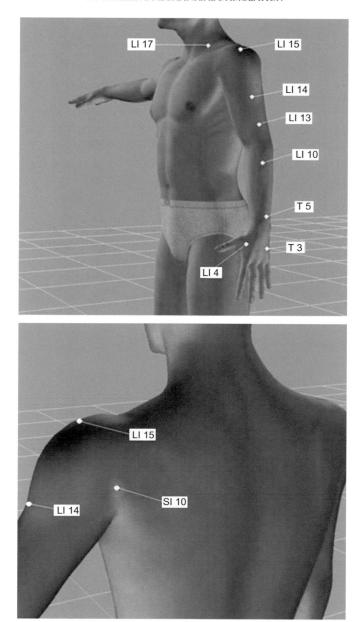

FIGURE 5.49 Recommended choice of APs for the treatment of brachialgia (always in addition to the upper back pain treatment).

5.8.1.3.9 C3 NEURALGIA

C3 neuralgia is usually a part of an upper back regional pain syndrome, but sometimes it may cause the main symptoms. Pressure applied to the C3 nerve roots may result in persistent pain not only in the posterior neck but also the ear region (especially with concomitant C2 nerve involvement), temporomandibular joint region, and even the frontal part of the neck (see Figure 5.35). Reflexive muscle spasm may involve not only posterior cervical and sternomastoid muscles but also superior constrictor and masticatory muscles, leading to limitation in the range of motion of the mandible and even swallowing difficulty. Prolonged reflexive vasoconstriction, with limited blood supply to the acoustic nerve, can be a cause of tinnitus.

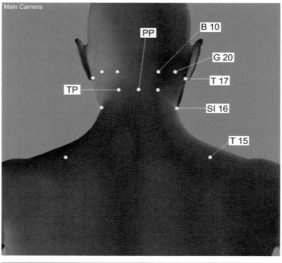
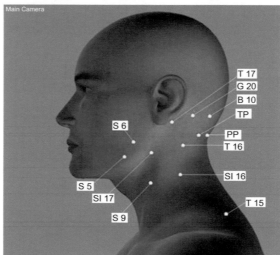
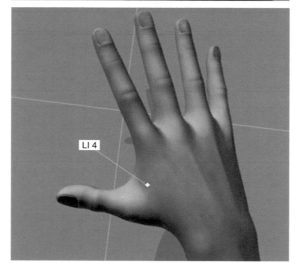

FIGURE 5.50 Recommended choice of classical APs and other PPs and TPs for the treatment of C3 neuralgia.

All of these symptoms can create diagnostic problems once all other potential problems in this region, such as otitis media, temporomandibular joint derangement, pharyngitis, etc., are ruled out. In our clinical material of dozens of patients with this particular pain syndrome, we have achieved an approximate 85% success rate with a complete or almost complete decrease in pain intensity on the analog pain scale and a complete or almost complete drop in the amount of medication used.

For practical reasons, the treatment of the C3 neuralgia should be done in a sitting position. Moderately strong palpation along the cervical spinal column will reveal increased sensitivity/tenderness between C2 and C3 spinous processes. Palpation along the C3 dermatome (see Figure 5.35) will localize the respective classical APs and other PPs/TPs ("Acute" or "Subacute" readings on the OED device) for stimulation. These points (Figure 5.50) should include the previously mentioned tender spot (PP) between the C2 and C3 spinous processes as well as TPs located at the same level, symmetrically on both sides of the neck, along the lines passing through classical APs B 9 and B 10. Among classical APs the most useful are: SI 16 (generally used for upper back pain; can be stimulated bilaterally with TENS), SI 17, T 16, T 17, G 20, B 10, S 5, S 6, and S 9. Because the C3 neuralgia is usually a part of regional pain syndrome, it is good also to stimulate T 15 (can be stimulated bilaterally with TENS) and traditionally, if tender, LI 4. If only one side is affected, the treatment can be applied unilaterally. If the other side is also affected, however (even slightly), the treatment should be applied symmetrically.

5.8.1.3.10 TRIGEMINAL NEURALGIA

It is still not clear what exactly causes this excruciating pain in distribution of one or more divisions of the trigeminal nerve, and it is very difficult to control with pharmacotherapy. Sometimes it can be the first signal of a serious systemic disease; therefore, a careful diagnosis is required. Beyond any doubt, when trigeminal neuralgia is present, there is a severe functional disturbance of the fifth cranial nerve's sensory component. Physical medicine should be, therefore, a leading therapy, due to its ability to cure the nerves. In our clinical material of dozens of patients with this particular pain syndrome, we have achieved an approximate 85% success rate with a complete or almost complete decrease in pain intensity on the analog pain scale and a complete or almost complete drop in the amount of medication used. However, the average full course of treatments was two to three times longer than the basic 20 therapeutic sessions.

Depending on which particular division is mostly affected, classical APs and other PPs ("Acute" or "Subacute" results on the OED device) located on the respective dermatome (see Figure 5.35) should be stimulated. In the case of mandibular division, the most useful APs are (Figure 5.51): T 17, T 20, S 4, S 5, S 6, S 7, G 3, G 8, and SI 18. An additional PP can also be tried, which is located symmetrically to the S 5 point on the internal side of the same corpus mandibulae (Figure 5.51). In the case of maxillary division (Figure 5.52): G 17, G 8, G 5, G 1, T 23, S 2, and LI 20. In the case of ophthalmic division (Figure 5.53): G 14, G 15, G 16, G 17, G 18, B 2, B 7, and S 1. If more than one division is involved, the previously mentioned points should be combined. Traditionally, LI 4 and S 44 can always be added as well as B 10, and G 20 can be tried (Figure 5.54) if tender ("Acute" or "Subacute" results on the OED device).

5.8.1.3.11 HEADACHE

Headache belongs to the most common of health problems, yet there are far too many people whose lives are ruined because of this condition. Headache can be a symptom of almost any disease; therefore, a proper diagnosis is of utmost importance. It seems that the OED screening can be tremendously useful in this regard by precisely estimating the origin and the intensity of the respective afferent nervous signals. Once the underlying problem is detected, the appropriate

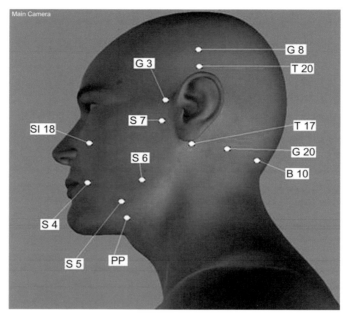

FIGURE 5.51 Recommended choice of classical APs as well as an additional PP for treatment of the mandibular division of the trigeminal nerve.

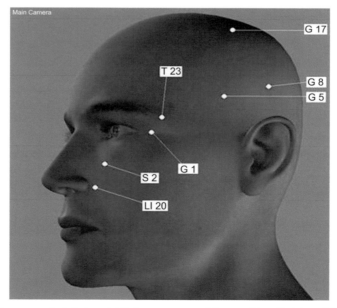

FIGURE 5.52 Recommended choice of APs for treatment of the maxillary division of the trigeminal nerve.

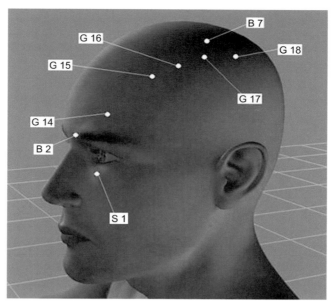

FIGURE 5.53 Recommended choice of APs for treatment of the ophthalmic division of the trigeminal nerve.

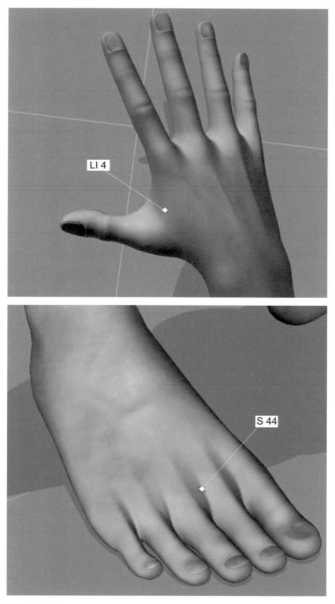

FIGURE 5.54 Supportive APs for the treatment of trigeminal neuralgia.

treatment should be introduced; fighting the symptoms with various pharmacotherapeutics does not make much sense.

There is a lot of confusion surrounding the pathophysiology of persistent headaches, with existing theories creating, in fact, more questions than answers; many of them concentrate more on technicalities than on clear, general ideas. Nevertheless, there are clinical indications that up to 85–90% of all headaches are caused by a mechanical pressure applied to the C1 and/or C2 nerve roots by solid spinal elements, such as cervical spondylosis (see Section 5.8.1.3.1). Therefore, some doctors/therapists rightfully use the term cervicogenic/cervical headache. The C1–2 region is relatively difficult for precise radiological assessment (the pain cannot be visualized!), but the OED will display an "Acute" result at the auricular OPA corresponding to the cervical spine. Physical examination will reveal increased sensitivity/tenderness above or/and below the level of the C1 spinous process. Palpation along the C2 dermatome (see Figure 5.35) will localize the respective classical APs and other related PPs ("Acute" or "Subacute" results on the OED device). This is because the pain "travels" along the affected nerve/nerves, producing the features of an occipital (migrainous) neuralgia. However, the pain automatically triggers muscle spasms (including neck muscles), leading to even more pressure applied to the pinched nerves (see Sections 3.3 and 5.1); during the acute phase this vicious circle can develop into a full clinical picture of the so-called cluster headache, characterized by a recurrent, severe, unilateral pain with radiation towards the eye. This kind of headache often occurs at night, perhaps due to the longer-lasting incorrect position of the head and neck. Once the problem becomes chronic, with muscles being continuously overcontracted and therefore sore on their own, the regional pain syndrome often described as a tension headache can be created. Mental stress can aggravate such a headache by strengthening muscle spasms, but it is not a primary cause!

When the problem becomes very serious, it can be called a migraine. Strong pressure on the C1 and/or C2 nerve roots can induce, in the reflexive way, a strong spasm of the respective muscles, including those small muscles in the cerebral arteries' walls. This will lead to vasoconstriction with reduction in the cerebral blood flow; limited oxygen supply will result in the so-called visual aura (blurred vision and flash points), often with dizziness and disorientation. Prolonged cerebral ischaemia may lead to rising intracranial pressure. The reduction in the cerebral blood supply will finally trigger the self-regulatory physiological mechanisms, which will increase the blood volume "pumped" to the brain, usually within the next 15 to 45 minutes. However, this increased cerebral and extracranial blood flow through still-narrowed arteries will raise an intra-arterial pressure, producing a characteristic throbbing headache. Severe "bursting" pain together with the reactive mental stress can induce nausea and vomiting. Higher estrogen levels in blood plasma, with subsequent water retention, contribute to a higher risk of migraines during the premenstrual period.

Because the pathogenesis of cervicogenic headache, including migraine, is directly connected with functioning of the nervous system, reflexive physical medicine, and in particular combined acupuncture therapy, is perfectly positioned when it comes to the effective treatment of this problem. In our clinical material of hundreds of patients with persistent severe headaches, we have achieved an approximate 85% success rate with a lasting complete or almost complete decrease in pain frequency and intensity on the analog pain scale and a complete or almost complete drop in the amount of medication used.

For practical reasons, the headache treatment should be done in a sitting position. Because the cervicogenic headache is usually a part of the upper back regional pain syndrome, it is good to start with bilateral stimulation (TENS) of SI 16 and T 15 (Figure 5.55). Combined acupuncture therapy should be applied to GV 15 (between C1 and C2 spinous processes) or GV 16 (above C1 spinous process), depending on which one displays more sensitivity to palpation (higher reading on the OED device percentage scale). Then all the PPs, including classical APs, located along the radiation of pain must be stimulated. The most useful APs (subject to tenderness and "Acute" or "Subacute" results on the OED device) are as follows: GV 20, B 10, B 9, T 17, G 20, G 19, G 18, G 16, G 14 (if strong frontal pain), and G 8. In each case, the so-called migraine PPs—located between the ears and eye brows—should also be stimulated. If the headache includes pain and muscle spasm in the temporomandibular region, the APs discussed in Section 5.8.1.3.9 should be added.

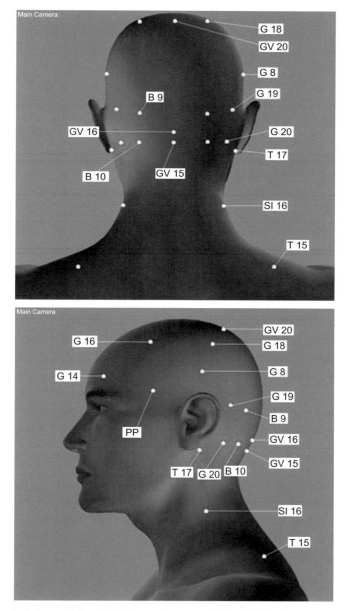

FIGURE 5.55 Recommended choice of classical APs as well as an additional PP for the treatment of cervicogenic headaches.

The treatments can be done on a daily basis during the acute phase and three times per week in the case of a chronic headache.

5.8.1.3.12 FIBROMYALGIA

It seems that the so-called fibromyalgia syndrome is one of the major misunderstandings in contemporary medicine. According to the American College of Rheumatology, the classification criteria for fibromyalgia syndrome (1990) consist of:

1. Widespread pain for three months or longer in an axial distribution plus pain on both sides of the body, as well as above and below the waist
2. The presence of 11 or more (recently, even six is considered enough) out of 18 specified tender points with moderate digital pressure of 4 kg (roughly the force required to blanch the examining nail when pressing against a hard surface).

There are also some associated symptoms that may include chronic pelvic pain, restless leg syndrome, irritable bowel syndrome and neurogenic bladder, noncardiac chest pain, persistent headache, dizziness, so-called chronic fatigue syndrome, and sleep disturbance. All of these symptoms are supposed to be caused by "central sensitization," with disordered sensory processing of pain impulses in the spinal cord. Apparently it is believed that there is a loss of pain regulation in the central nervous system that causes pain amplification.

However, all of these diagnostic criteria and symptoms are, first of all, typical for a widespread chronic PSSO (see Section 5.8.1.3.1). Pinched nerve roots at a certain spinal level can be a cause of the pain/muscle spasm/pain vicious circle, which in sensitive people (those with a lower pain threshold?) can result in vast chronic muscle spasms, producing all the previously mentioned symptoms. All of the 18 specified tender points belong to known APs used for the treatment of spinal problems. Even certain chemical changes observed in the respective muscles are typical for the prolonged muscle overcontraction and cannot be used as an evidence of fibromyalgia.

We have successfully treated hundreds of patients who fully qualified for the fibromyalgia syndrome diagnosis, many of them with no visible abnormalities on spinal X-rays/scans. Nevertheless, in each case the OED device signaled an "Acute" result at the auricular areas corresponding to the respective levels of the spine. Also, moderately strong palpation in each case revealed PPs/TPs ("Acute" or "Subacute" readings on the OED device) at the relevant spinal levels and along related dermatomes. Therefore, all these patients have been considered as PSSO and treated accordingly; combined acupuncture therapy appeared to be of utmost usefulness.

The main problem of the fibromyalgia syndrome diagnosis seems to be that the pain cannot be visualized and therefore cannot be precisely traced and assessed with contemporary imaging diagnostic facilities such as MRI, CT scan, X-ray, or ultrasound examination. However, even the doctors/therapists with no access to the OED technology can make the right diagnosis based on a properly performed physical examination of the spinal column. Making an artificial diagnosis of fibromyalgia will deprive PSSO patients of the best therapeutic option, which is arguably reflexive physical medicine.

Therefore, the management of patients referred with fibromyalgia syndrome should start with a moderately strong palpation all along the spinal column to reveal a presence of very tender spots between the spinous processes; in this way, the affected levels and respective dermatomes will be precisely identified. Depending on the outcome, the procedures described in the respective "Spinal" and "Headache" sections should be performed.

Unless proven wrong, at this stage we seriously doubt the existence of fibromyalgia as a separate disease or disorder on its own. Instead, we presume this syndrome to be misdiagnosed PSSO. Even the term "fibromyalgia" seems unfortunate; "myalgia" alone would be acceptable (due to persistent spasm), but what for "fibro"? From a practical point of view, the most important point is that reflexive physical medicine is still properly used in all these cases as the first therapeutic choice.

5.8.1.3.13 MYOFASCIAL PAIN

Typical myofascial pain results from the strain or improper use of a muscle. It usually starts and stops suddenly. Contrary to PSSO, pain is worse at rest than during exercise. Nevertheless, careful differential diagnosis is required between these two conditions; OED can be very helpful.

Reflexive physical medicine, and especially combined acupuncture therapy, can still be a leading therapeutic option, in particular when medication (anti-inflammatories, painkillers and rather useless antiepileptics and antidepressants) as well as therapeutic exercises do not help sufficiently. First of all, the local PPs/TPs should be stimulated ("Acute" or "Subacute" result on the OED device), as well as those located along the respective dermatome. As already mentioned in Section 5.8.1.2.3, the treatments can be done on a daily basis during the acute phase and three times per week in case of a chronic pain.

5.8.1.3.14 PHANTOM PAIN

Phantom pain is a pain sensation felt in a body part that has been removed. Historically, neuromas formed from injured nerve endings at the stump site were thought to be the main cause of the phantom pain. However, blocking of the respective peripheral nerves' conduction usually does not stop the sensation; this indicates a different origin of the phantom pain. It seems that the leading pathophysiological mechanism might be still the one of sciatica, brachialgia, or intercostal neuralgia (in case of mastectomy), with persistent physical pressure applied to the nerves entering the spinal cord by prolapsed intervertebral discs and/or degeneratively deformed vertebral elements.

Therefore, in every case of phantom pain, a proper diagnosis is required; the OED can be very useful in identifying the source of pain. Patients with possible underlying PSSO should be treated accordingly; see Sections 5.8.1.3.2, 5.8.1.3.7, and 5.8.1.3.8. Naturally, all the local PPs present at the stump site ("Acute" or "Subacute" readings on the OED device) must always be stimulated. Treatments can be done on a daily basis during the acute phase and three times per week in the case of a chronic pain. Phantom pains can be very persistent and therefore might require a prolonged course of therapeutic sessions.

5.8.1.3.15 PAINFUL SHOULDER

Shoulder pain can be caused by variety of pathogenic factors; therefore, a proper diagnosis is required. Interestingly, in a majority of cases, the shoulder region pain is caused by pinched roots of the brachial nerve (see Section 5.8.1.3.8); the OED allows a rapid differential diagnosis. Doctors/therapists without access to OED technology should always examine the patient's cervical spinal column with moderately strong palpation in search of very tender spots between spinous processes C3–Th 3; the presence of such PPs indicates a cervicogenic origin of the problem, especially when there is no convincing X-ray or ultrasound evidence of local shoulder pathology.

In proven local shoulder problems that do not require surgical intervention, the use of physical medicine and especially combined acupuncture therapy is indicated for pain control, regional muscle spasm relaxation, and reduction of inflammation and infection through the improved local blood supply. In the case of rheumatoid arthritis and other autoimmunological diseases, physical medicine can be still helpful, as a supportive symptomatic therapy; improved local blood supply will not increase the amount of antibodies reaching the shoulder joint: There are already enough of them.

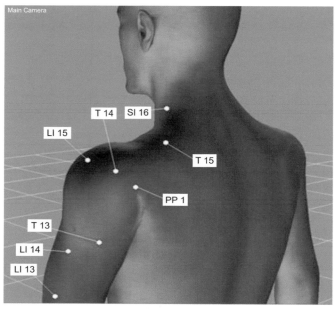

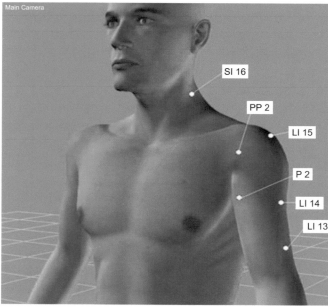

FIGURE 5.56 Recommended choice of classical APs and other PPs (PP 1 and PP 2) for the treatment of a painful shoulder.

Because a painful shoulder causes, in the reflexive way, a regional muscle spasm, it is good to start the treatment with bilateral stimulation (TENS) of SI 16 and T 15 (Figure 5.56). Other recommended APs (subject to tenderness and "Acute" or "Subacute" readings on the OED device) are, as follows: LI 13, LI 14, LI 15, T 13, T 14, and P 2. Most important, however, are two PPs located symmetrically on the anterior and posterior aspects of the shoulder (Figure 5.56); it is practical to stimulate them with TENS. Depending on the location of pain, the treatment can be done unilaterally or bilaterally. Treatments can be done on a daily basis during the acute phase and three times per week in the case of a chronic pain.

5.8.1.3.16 PAINFUL ELBOW

Elbow pain can be a clinical manifestation of certain systemic diseases, such as polyarthritis of various origins, but in such a case the other joints should also be affected. If the elbow is the only joint affected and there is no obvious surgical problem, then usually the pain is caused by overuse or strain during sports or other physical activities: so-called tennis elbow.

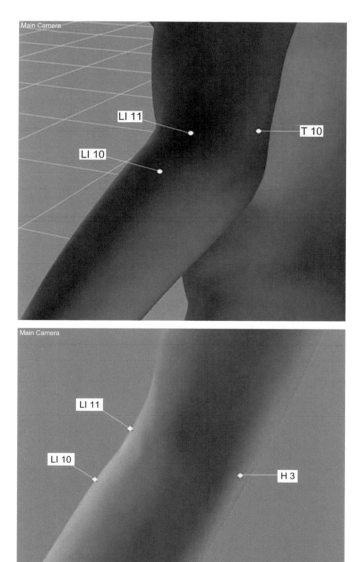

FIGURE 5.57 Recommended choice of APs for the treatment of a painful elbow.

Physical medicine, and in particular combined acupuncture therapy, can be very helpful in the management of tennis elbow and other painful elbow problems. Even in the case of rheumatoid arthritis and other autoimmunological diseases, physical medicine can be still used as a supportive symptomatic therapy; improved local blood supply will not increase the amount of antibodies reaching the elbow joint: There are already enough of them. For pain and inflammation control, the stimulation of the following APs (Figure 5.57) is recommended (subject to tenderness and "Acute" or "Subacute" readings on the OED device): LI 10, LI 11, H 3, and T 10. Treatments can be done on a daily basis during the acute phase and three times per week in the case of a chronic pain. Patients with tennis elbow should stop their physical activities until the pain is completely gone.

5.8.1.3.17 PAINFUL WRIST/CARPAL TUNNEL SYNDROME

Wrist pain can be a clinical manifestation of certain systemic diseases, such as polyarthritis of various origins, but in such a case the other joints are also affected. If the wrist is the only joint

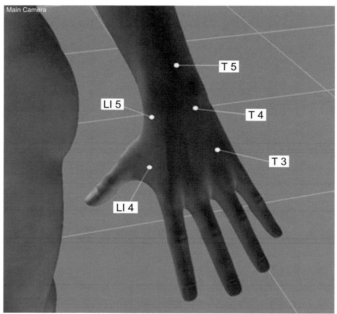

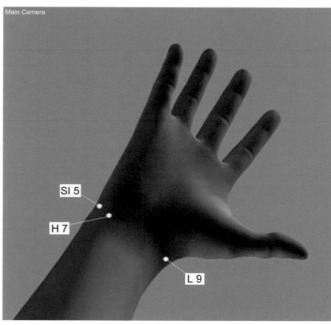

FIGURE 5.58 Recommended choice of APs for the treatment of a painful wrist or carpal tunnel syndrome.

affected and there is no evidence of fracture or other obvious surgical problems, then the pain is probably caused by a soft tissue injury or a course of microinjuries sustained during sports or other physical activities. Soft tissue swelling of any origin may cause median nerve compression within the unyielding carpus, producing symptoms of so-called carpal tunnel syndrome. Typical symptoms can be similar to those of brachialgia and include wasting, tingling, and burning pain in the hand, especially at night; careful differential diagnosis is required, and the OED can be very helpful.

Physical medicine, and in particular combined acupuncture therapy, can be successfully used for the control of wrist pain, inflammation, and soft tissue swelling. Even in the case of rheumatoid arthritis and other autoimmunological diseases, physical medicine can be still used as a supportive symptomatic therapy; improved local blood supply will not increase the amount of antibodies reaching the wrist joint: There are already enough of them.

For treatment of wrist pain, inflammation, and swelling, the following APs (Figure 5.58) can be stimulated if tender ("Acute" or "Subacute" readings on the OED device): L 9, LI 5, T 4, T 5, SI 5, and H 7. In the case of proven carpal tunnel syndrome, LI 4 and T 3 can be added (Figure 5.58). Treatments can be done on a daily basis during the acute phase and three times per week in the case of a chronic pain.

5.8.1.3.18 PAINFUL HIP

In the majority of cases, it is sciatica that causes pain in the hip region (see Section 5.8.1.3.2). Therefore, careful differential diagnostics is required, especially when there is radiological evidence of both lumbo-sacral and local hip pathology; in such a case, the OED can clearly identify the primary source of pain.

In proven nonsurgical, local hip problems such as osteoarthritis, physical medicine should be still tried before any invasive procedure such as a hip replacement is considered. Combined acupuncture therapy might be especially helpful, even in the case of prolonged pain after the replacement. Rheumatoid arthritis and other autoimmunological diseases are not direct contraindications for the use of physical medicine as a supportive therapy; improved local blood supply can be very beneficial, and it will not increase the amount of antibodies reaching the hip joint: There are already enough of them.

In the management of a painful hip region, the use of following APs (Figure 5.59) should be considered (subject to tenderness and "Acute" or "Subacute" readings on the OED device): G 28, G 29, G 30, G 31, SP 11, SP 13, and S 30. Treatments can be done on a daily basis during the acute phase and three times per week in the case of a chronic pain.

5.8.1.3.19 PAINFUL KNEE

Knee pain can be a clinical manifestation of certain systemic diseases, such as polyarthritis of various origins, but in such a case the other joints are also affected. If the knee is the only joint affected, then the pain is probably caused by a soft tissue injury or a course of microinjuries sustained during sports or other physical activities.

Physical medicine, and in particular combined acupuncture therapy, can be very useful in the management of all nonsurgical painful knee conditions. Even in the case of rheumatoid arthritis and other autoimmunological diseases, physical medicine can be still used as a supportive symptomatic therapy; improved local blood supply will not increase the amount of antibodies reaching the knee joint: There are already enough of them. The most useful classical APs are B 40 and G 33. However, equally recommended is the stimulation of all the other PPs ("Acute" or "Subacute" readings on the OED device) found in the area; three of them are marked on Figure 5.60. The first one (PP 1), the most important "specific knee point," is located just below the knee joint, at the antero-internal aspect of the lower leg. The second one (PP 2) lies on the same vertical line, but just above the knee. The third one (PP 3) is located on the SP meridian, above the classical AP SP 9: approximately 1/3 of the distance between SP 9 and SP 10. Treatments can be done on a daily basis during the acute phase and three times per week in the case of a chronic pain.

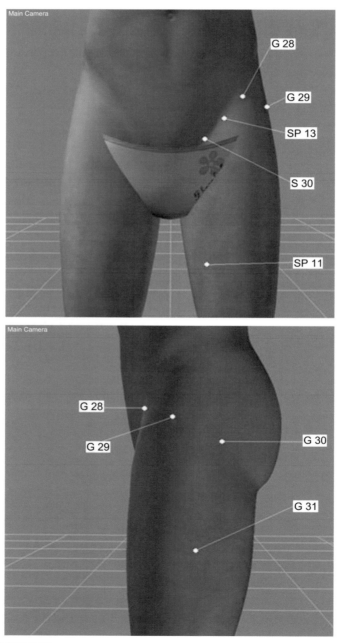

FIGURE 5.59 Recommended choice of APs for the treatment of a painful hip.

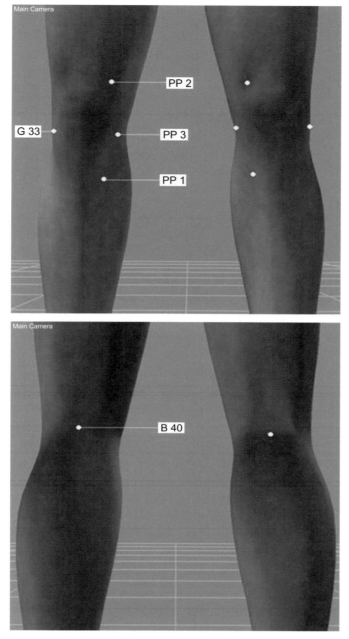

FIGURE 5.60 Recommended choice of classical APs and other PPs for the treatment of a painful knee.

5.8.1.3.20 PAINFUL ANKLE

Ankle pain can be a clinical manifestation of certain systemic diseases, such as polyarthritis of various origins, but in such a case the other joints are also affected. If the ankle is the only joint affected, then the pain is probably caused by a soft tissue injury or a course of microinjuries sustained during sports or other physical activities.

Physical medicine, and in particular combined acupuncture therapy, can be very useful in the management of all nonsurgical painful ankle conditions. Even in the case of rheumatoid arthritis and other autoimmunological diseases, physical medicine can be still used as a supportive symptomatic therapy; improved local blood supply will not increase the amount of antibodies reaching the knee joint: There are already enough of them.

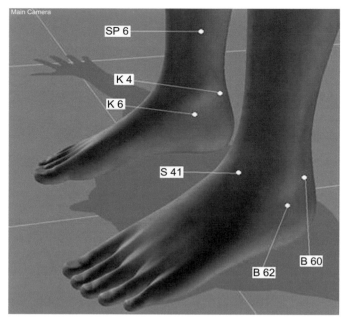

FIGURE 5.61 Recommended choice of APs for the treatment of a painful ankle.

In the management of a painful ankle region, the use of following APs (Figure 5.61) should be considered (subject to tenderness and "Acute" or "Subacute" readings on the OED device): B 60, B 62, S 41, K 4, K 6, and SP 6. Treatments can be done on a daily basis during the acute phase and three times per week in the case of a chronic pain.

5.8.1.3.21 DIABETIC NEUROPATHY

Although diabetic neuropathy is rarely a direct cause of death, it can become a major source of suffering. The most common clinical picture is that of peripheral polyneuropathy. Usually bilateral, the symptoms include numbness, paresthesias, severe hyperaesthesias, and pain, especially in the plantar areas of both feet. The pain is typically worst at night and may be deep-seated and severe, often lancinating or lightning in type. If not dealt with correctly, the condition can lead to foot ulcers and ultimately amputation of the feet. Unfortunately, current pharmacotherapy of diabetic neuropathy (analgesics, NSAIDs, antiepileptics, and antidepressants) is unsatisfactory in most respects.

Because physical medicine is very well positioned when it comes to the regeneration of severely damaged nerves (see Sections 3.3 and 5.1), we have conducted a pilot study (28) followed by a randomized, placebo-controlled full clinical study (66) to estimate the usefulness of combined acupuncture therapy in the management of diabetic neuropathy. The studies confirmed that this type of treatment is very effective, safe, convenient, and affordable, and leads to lasting clinical improvement, not only pain control. Therefore, we believe that combined acupuncture therapy should be used as a leading treatment in each case of diabetic neuropathy. There are ongoing discussions in international medical literature and at scientific medical congresses of how to treat this "incurable" condition in a chemical way; meanwhile, physical medicine simply does the job in a spectacular manner!

The most important classical APs to be considered for the treatment of diabetic neuropathy, subject to tenderness and "Acute"/"Subacute" readings on the OED device, are as follows (Figure 5.62): SP 6, SP 4, K 4, K 6, B 58, B 60, B 62, B 63, S 36, S 43, and G 34 (all of them to be

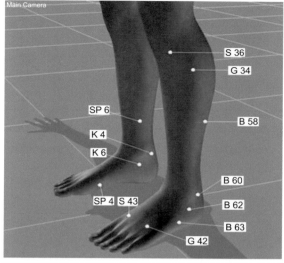
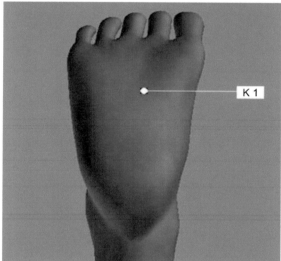
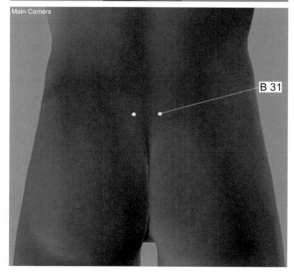

FIGURE 5.62 Recommended choice of APs for the treatment of diabetic neuropathy.

used bilaterally); K 1 and G 42 can be bilaterally stimulated with TENS (same electric channel for each foot) and cryotherapy for practical reasons. If the lower back pain coexists (this occurs very often, especially in elderly patients), a simultaneous stimulation of APs described in Section 5.8.1.3.2 should be considered. If not, it is still good to stimulate bilaterally an AP B 31, located at the first posterior sacral foramen. The treatments should be done at least three times per week. The full course of treatments usually takes much more than the basic twenty therapeutic sessions; the number of sessions required is very individual and depends on the severity of the condition. It is advisable that patients continue their TENS treatments (see Section 5.6.1 and Figure 5.9) on a regular basis at home, even after the successful course of combined acupuncture therapy is completed.

5.8.1.3.22 RAYNAUD'S SYNDROME

This is an intermittent vasospasm of the arterioles of the distal limbs after exposure to cold or emotional stimuli. There is usually a temporary color change (pale/blue/red) and pain in the affected part. It may be idiopathic (Raynaud's disease), but may also be secondary to an underlying cause, such as trauma, collagenoses, thoracic outlet obstruction, atherosclerosis, or cryoglobulins; therefore, a careful diagnosis is required. Current pharmacotherapy is rather unsatisfactory, as are the results of more radical measures, such as sympathectomy.

Physical medicine is well positioned when it comes to the treatment of functional vascular disturbances (see Sections 3.3 and 5.1), and the course of combined acupuncture therapy can result in lasting clinical improvement. The most useful classical APs to be considered for the treatment of Raynaud's syndrome, subject to tenderness and "Acute"/"Subacute" readings on the OED device, are as follows (Figure 5.63): T 15 (can be stimulated bilaterally with TENS), T 5, T 3, LI 10, LI 4, and P 6. The treatments should be done at least three times per week.

5.8.1.3.23. RHINITIS

Through improved local blood circulation, physical medicine, and in particular combined acupuncture therapy, can be very helpful in controlling the nasal mucous membrane inflammation with edema and extensive rhinorrhea. In allergic, atrophic, or vasomotor rhinitis, physical medicine can be used as a leading or the only therapy aimed at lasting clinical improvement; in our clinical material of dozens of rhinitis patients, we have achieved an approximate 80% success rate with a complete disappearance of all symptoms and a complete or almost complete drop in the amount of medication used. In the case of chronic rhinitis caused by other known pathogenic factors, physical medicine can be still used as a supportive symptomatic treatment.

The most useful classical APs to be stimulated bilaterally in the case of rhinitis, subject to tenderness and "Acute"/"Subacute" readings on the OED device, are as follows (Figure 5.63): LI 4, LI 11, LI 20, L 7, B 2, GV 22, G 20, S 2, and Liv 3. An additional (nonclassical) AP called Yintang may also be used; it is located on the Governing Vessel (GV), between the eyebrows. In the case of severe symptoms, it is good to add two PPs ("Acute" or "Subacute" result on the OED device) located symmetrically on the outer sides of nostrils on the verge of nasal bones (Figure 5.64). The treatments should be done at least three times per week.

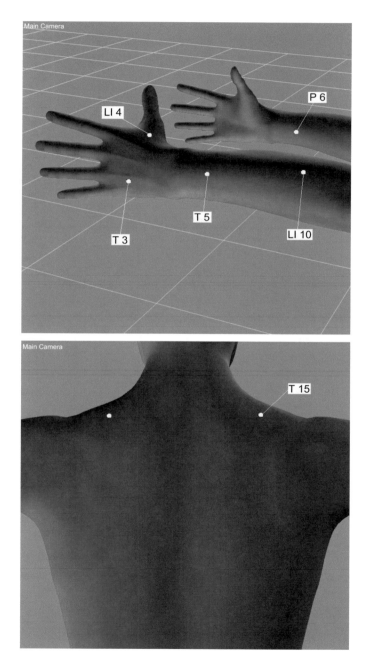

FIGURE 5.63 Recommended choice of APs for the treatment of Raynaud's syndrome.

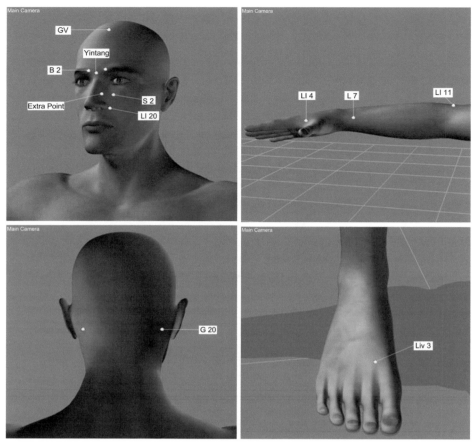

FIGURE 5.64 Recommended choice of classical APs and other PPs for the treatment of rhinitis.

5.8.1.3.24 SINUSITIS

During the course of rhinitis, the swollen nasal mucous membrane obstructs the ostium of the paranasal sinus. The oxygen in the sinus is absorbed into the blood vessels in the sinus mucous membrane. In this way, a relative negative pressure in the sinus (vacuum sinusitis) is created; a transudate from the mucous membrane develops and fills the sinus, where it serves as a medium for bacteria and other germs. An outpouring of serum and leucocytes to combat the infection results, and painful positive pressure develops in the obstructed sinus. Thus, a viscious circle takes place: the more intensive the inflammation, the more hyperaemic and edematous becomes the mucous membrane "sealing" the sinus obstruction and creating the ideal conditions for chronic infection.

Through better local blood circulation, physical medicine, and in particular combined acupuncture therapy, can reduce edema of the mucous membrane and in this way improve drainage of the affected sinus. Better local blood supply will also help to control the infection. Even medication will have now much better access to the infected area. Therefore, physical medicine should always be considered when dealing with sinusitis; in our clinical material of dozens of patients with chronic sinusitis, we have achieved an approximate 80% success rate with a complete

FIGURE 5.65 The APs G 14 for the treatment of sinusitis, always in addition to all the APs recommended for the treatment of rhinitis

disappearance of all symptoms and a complete or almost complete drop in the amount of medication used.

All the APs recommended for the treatment of rhinitis (see Section 5.8.1.3.23) should be also used for sinusitis, with bilateral addition of G 14 (Figure 5.65), subject to tenderness and "Acute"/"Subacute" readings on the OED device. In the case of chronic sinusitis, the treatments should be done at least three times per week.

5.8.1.3.25 HEARING LOSS DUE TO ACOUSTIC NERVE DAMAGE/TINNITUS

Through improved local blood circulation, physical medicine, and in particular combined acupuncture therapy, is well positioned when it comes to the regeneration of severely damaged nerves (see Sections 3.3 and 5.1). Therefore, it should always be tried in the case of tinnitus and hearing loss due to acoustic nerve damage. In our clinical material of several patients with severe hearing loss of this type, in each case we managed to improve the condition to a level allowing free communication with no need for the use of a hearing aid; the audiograms confirmed significant clinical improvement. However, the average full course of treatments was at least three times longer than the basic twenty therapeutic sessions. In our clinical material of dozens of patients with tinnitus, we have achieved an approximate 75% success rate.

The most useful APs to be stimulated in the case of a hearing loss due to the acoustic nerve damage, subject to tenderness and "Acute"/"Subacute" readings on the OED device, are as follows (Figure 5.66): T 3, T 17, G 7, G 11, G 20, and SI 19. The same APs can also be used for the treatment of tinnitus. Therapeutic sessions should be done at least three times per week.

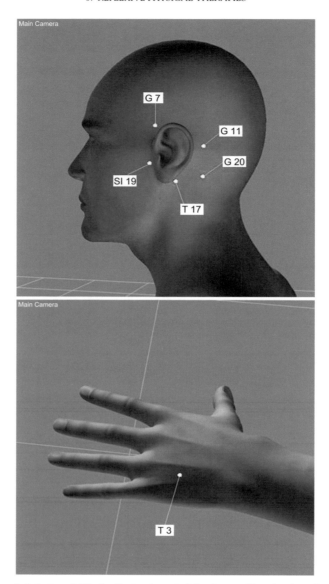

FIGURE 5.66   Recommended choice of APs for the treatment of tinnitus and hearing loss due to acoustic nerve damage.

5.8.1.3.26 MÉNIÈRE'S DISEASE

This is a disorder characterized by recurrent prostrating vertigo, sensory hearing loss, and tinnitus. It is associated with generalized dilation of the membranous labyrinth (endolymphatic hydropsy). As is the case with many other functional disorders, Ménière's disease is an indication for combined acupuncture therapy aimed at both breaking the current attacks and lasting clinical improvement.

Because Ménière's disease causes, in the reflexive way, a regional muscle spasm, it is good to start the treatment with bilateral stimulation (TENS) of SI 16 and T 15. Other recommended APs to be stimulated only on the affected side, subject to tenderness and "Acute"/"Subacute" readings on the OED device, are as follows (Figure 5.67): K 7, T 3, T 17, G 7, G 11, G 20, SI 19, B 10, and GV 20. If possible, treatments should be done whenever the attacks occur; otherwise, they should take place at least three times per week.

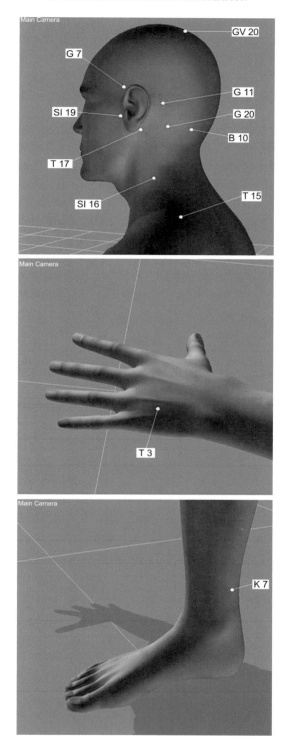

FIGURE 5.67 Recommended choice of APs for the treatment of Ménière's disease.

5.8.1.3.27 BRONCHIAL ASTHMA

This is a generalized airways obstruction, which can be a life-threatening condition, but, in the early stage at least, it is paroxysmal and reversible. All three major obstructing factors, bronchial muscle constriction, mucosal swelling, and increased mucus production, can be well controlled by physical medicine, and in particular by combined acupuncture therapy, but the most spectacular and rapid improvement concerns the bronchospasm. Bronchospasm is a result of activation of the local reflex arc by an inflammatory allergic reaction in the bronchial mucous membrane. Acupuncture, along with other reflexive therapies, can block the respective afferent signals and in this way stop the reflex arc (see Section 3.0). Therefore, combined acupuncture therapy can be used both as a life-rescuing procedure, able to break even a status asthmaticus, and as a maintenance treatment aimed at obtaining long-lasting clinical improvement.

In treating dozens of serious bronchial asthma cases, we have witnessed dramatic changes in patients' clinical conditions after each acupuncture treatment, with spirometric parameters rapidly improving (unpublished observations). Intensive courses of combined acupuncture therapy should be considered as an important supportive treatment in all severe cases of bronchial asthma undergoing hospitalization. Less severe cases would benefit much from the combined acupuncture courses performed on a less frequent, but still regular, outpatient basis; in this way, the amount of medication used could be significantly reduced and sometimes even completely stopped.

The most useful APs to be stimulated bilaterally in the case of bronchial asthma, subject to tenderness and "Acute"/"Subacute" readings on the OED device, are as follows (Figure 5.68): LI 4,

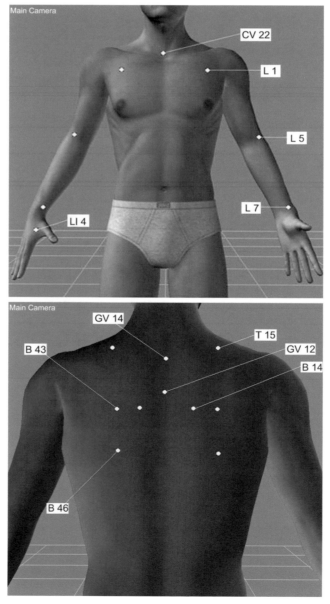

FIGURE 5.68 Recommended choice of APs for the treatment of bronchial asthma.

L 7, L 5, L 1, CV 22, T 15 (can be stimulated with TENS), GV 14, GV 12, B 14, B 43, B 46 (can be stimulated with TENS), and S 40. If possible, treatments should be done whenever the dyspnea starts; otherwise, they should take place on a daily basis during the acute phase and at least three times per week once the crisis is over.

5.8.1.3.28 ESOPHAGOSPASM/ACHALASIA

Combined acupuncture therapy is proven to be very helpful in the management of a failure of esophageal peristalsis and a failure of relaxation of the lower esophageal sphincter. It can be used as an emergency treatment to prevent dysphagia, but the complete course of treatments should result in a full recovery.

The most useful APs to be stimulated in the case of esophagospasm, subject to tenderness and "Acute"/"Subacute" readings on the OED device, are as follows (Figure 5.69): CV 17, CV 15, CV 12, K 26, B 15, and B17 (the last three bilaterally). If possible, treatments should be done whenever the spasm occurs; otherwise, they should take place on a daily basis during the acute phase and then at least three times per week.

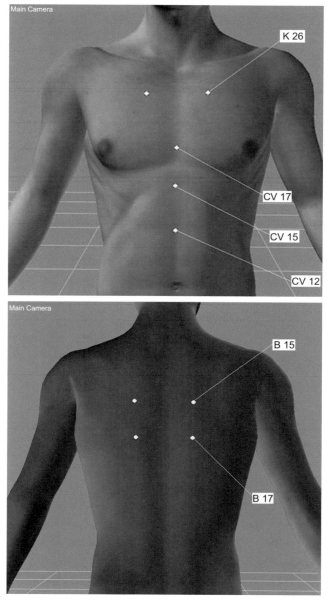

FIGURE 5.69 Recommended choice of APs for the treatment of esophagospasm/achalasia.

5.8.1.3.29 GLOBUS HYSTERICUS

Anxiety may sometimes produce the experience of a lump in the throat at times when the patient is not swallowing: globus hystericus. Of course, to make this diagnosis, all other serious possibilities have to be ruled out. Anxiolytics and psychotherapy are not always helpful in the case of this unpleasant condition.

The course of combined acupuncture therapy can lead to the lasting improvement. The most recommended APs, subject to tenderness and "Acute"/"Subacute" readings on the OED device, are as follows (Figure 5.70): T 15 and SI 16 (both can be stimulated with TENS), LI 4, P 6, G 20, S 9 (all of them bilaterally), CV 22, and CV 23. Because globus hystericus is usually a chronic condition, the treatments should be done at least three times per week until the condition is completely cured.

5.8.1.3.30 ULCERATIVE COLITIS

This idiopathic inflammatory disease of the large intestine can be difficult to control with medication; therefore, on many occasions a colectomy is necessary. Combined acupuncture therapy can be a valuable supportive treatment, helpful in the acute phase of the disease and for preventing relapses.

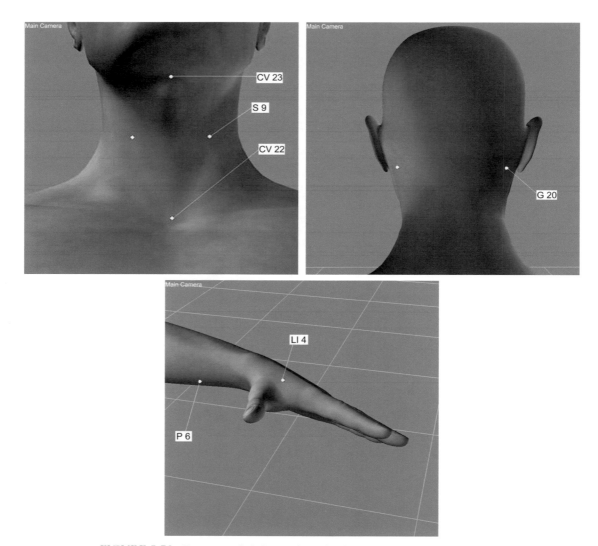

FIGURE 5.70 Recommended choice of APs for the treatment of globus hystericus.

The most useful APs to be stimulated bilaterally in the case of ulcerative colitis, subject to tenderness and "Acute"/"Subacute" readings on the OED device, are as follows (Figure 5.71): S 36, S 21, SP 9, SP 15, CV 5, CV 12, and K 3. Treatments can be done on a daily basis during the acute phase and then at least three times per week until a satisfactory clinical improvement is achieved.

5.8.1.3.31 RENAL COLIC

Renal colic, which produces one of the worst imaginable pains, can be a good example of both the effectiveness of acupuncture therapy and an explanation of its mode of action. As already mentioned in Section 5.1, the stone traversing the ureter hurts the ureter from inside and in this

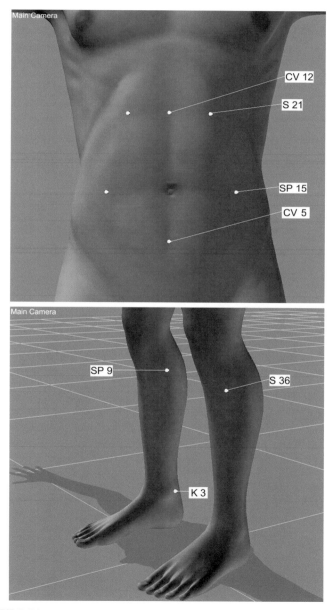

FIGURE 5.71 Recommended choice of APs for the treatment of ulcerative colitis.

way triggers the local reflex arc, which finally results in the strong contraction of the ureteric muscles and "imprisonment" of the stone. The stone firmly blocks the urine flow, causing the kidney to swell like a balloon (hydronephrosis); this directly produces extremely painful renal colic. A vicious circle is created: the stronger the urine pressure on the stone, the stronger the muscle spasm and the urine blockage.

Acupuncture, along with other reflexive therapies, can block the respective afferent signals and in this way stop the reflex arc (see Section 3.0). This will result not only in spectacular pain blockage, but more importantly in relaxation of the ureteric muscle contraction, allowing the stone to pass smoothly to the urinary bladder. This means that in this particular case even the cause of suffering can be removed by means of physical therapy, as long the size of the stone is not too big.

The most useful APs to be stimulated unilaterally on the side of renal colic, subject to tenderness and "Acute"/"Subacute" readings on the OED device, are as follows (Figure 5.72): B 23, G 25, G 26, G 27, S 30, SP 6, K 2, and K 5. Because the patient undergoing the treatment is usually lying on his or her back, it is practical to stimulate the APs B 23 and S 30 with TENS. If possible, the combined acupuncture therapy should be applied whenever the pain starts and should be continued until the pain is finished. Otherwise, during the pain-free intervals, the treatments can be done on a daily basis. A simultaneous use of medication can accelerate the final outcome, even if medicine alone was unable to produce relief.

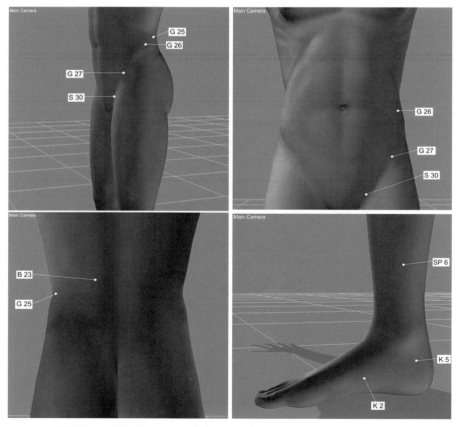

FIGURE 5.72 Recommended choice of APs for treatment of renal colic.

5.8.1.3.32 NEUROGENIC/ATONIC BLADDER

Functional disturbances of the urinary bladder often accompany serious lumbo-sacral spinal problems (see Section 5.8.1.3.1); in such a case, proper treatment of the spinal problem should also improve bladder function. Otherwise, if all other kinds of primary pathology are ruled out, the

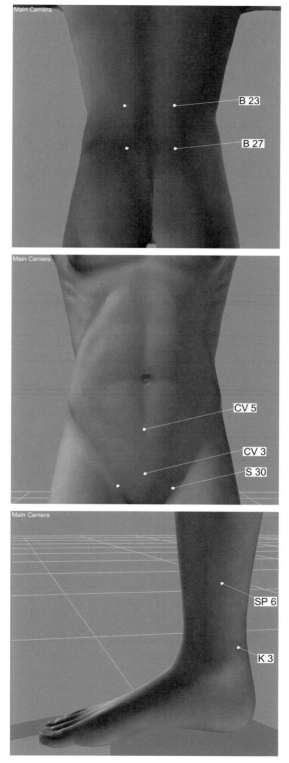

FIGURE 5.73 Recommended choice of APs for treatment of neurogenic/atonic bladder.

so-called neurogenic bladder can be successfully addressed with physical medicine, and especially combined acupuncture therapy. The so-called atonic bladder can also be addressed this way, after surgical abdominal intervention.

In order to restore normal bladder function, the following APs should be stimulated (Figure 5.73), subject to tenderness and "Acute"/"Subacute" readings on the OED device: B 23 and B 27 (for practical reasons, both can be stimulated bilaterally with TENS), K 3, SP 6, S 30, CV 3, and CV 5. In the case of atonic bladder, treatments can be done on a daily basis until the normal bladder function is restored; in the case of neurogenic bladder, treatments should be performed three times per week.

5.8.1.3.33 NOCTURNAL ENURESIS

This stressful condition requires careful diagnostic analysis to rule out any clear pathological factors. However, the majority of cases present with functional (psychogenic?) enuresis, and in these cases combined acupuncture therapy can be the most effective therapeutic option.

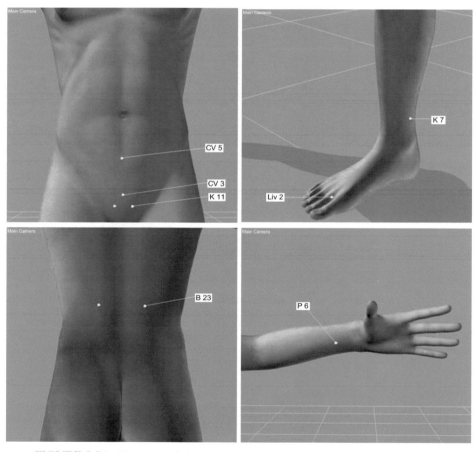

FIGURE 5.74 Recommended choice of APs for the treatment of nocturnal enuresis.

The most useful APs to be stimulated in the case of nocturnal enuresis, subject to tenderness and "Acute"/"Subacute" readings on the OED device, are as follows (Figure 5.74): B 23 (for practical purposes, can be stimulated bilaterally with TENS), Liv 2, K 7, K 11, CV 3, CV 5, and P 6. Treatments can be done three times per week; they should be stopped once improvement occurs (two to three subsequent nights without enuresis).

5.8.1.3.34 HYPEREMESIS GRAVIDARUM

Excessive vomiting during pregnancy may create a serious medical problem. Combined acupuncture therapy is proven to be very helpful in the management of hyperemesis gravidarum, but the use of this radical treatment has to be carefully considered due to the potential risk of abortion; stimulation of APs located on the legs, lower back, and lower abdomen is generally contraindicated in pregnancy.

If a combined acupuncture therapy is to be used, however, then the following combination of APs (Figure 5.75) is the most recommended, subject to tenderness and "Acute"/"Subacute" readings on the OED device: CV 14, P 6, S 36, and S 9. If possible, treatments should be done every time before the vomiting starts; otherwise, they can be performed on a daily basis during the acute phase and then at least three times per week as long there is a risk of excessive vomiting.

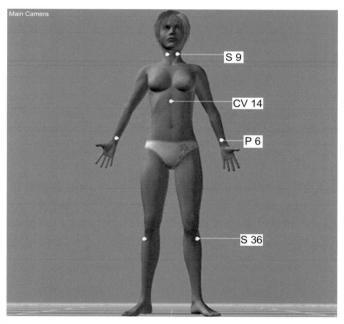

FIGURE 5.75 Recommended choice of APs for the treatment of hyperemesis gravidarum.

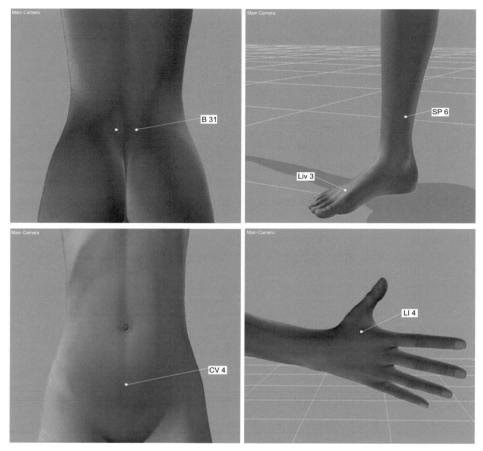

FIGURE 5.76 Recommended choice of APs for induction of labor uteral contractions.

5.8.1.3.35 DELAYED DELIVERY

Combined acupuncture therapy can be still tried before a Cesarean section in the case of postmature delivery, when oxytocin and prostaglandin drips did not induce uteral contractions. We have treated several cases of this kind; in each case, electroacupuncture combined with TENS induced strong uteral contractions within 30 to 45 minutes. The average time of such an induced labor was shorter than the typical one; there was also an analgetic effect of the combined acupuncture treatment observed.

The most useful APs to be stimulated in the case of delayed delivery, subject to tenderness and "Acute"/"Subacute" readings on the OED device, are as follows (Figure 5.76): B 31 (for practical reasons can be bilaterally stimulated with TENS), SP 6, Liv 3, LI 4, and CV 4. In order to maintain strong uteral contractions, both electroacupuncture and TENS can be continued over the period of many hours.

5.8.3.1.36 RETINOPATHIES/MACULAR DEGENERATION/OPTIC NERVE ATROPHY/GLAUCOMA

Physical medicine, and especially combined acupuncture therapy, can significantly improve local blood circulation. Therefore, this kind of treatment should be considered in the management of certain ophthalmologic diseases or disorders that might be related to disturbances in the local blood circulation. After all, contemporary ophthalmology has not much to offer in majority of these problems!

The most useful APs to be stimulated in the case of blood-supply related ophthalmologic diseases or disorders, subject to tenderness and "Acute"/"Subacute" readings on the OED device, are as follows (Figure 5.77): G 1, G 14, G 20, G 37, G 43, B 2, B 10, S 2, Liv 3, and K 6. In acute cases,

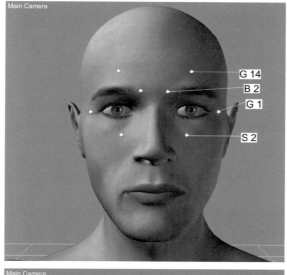
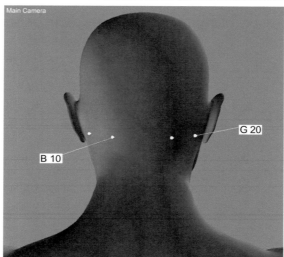
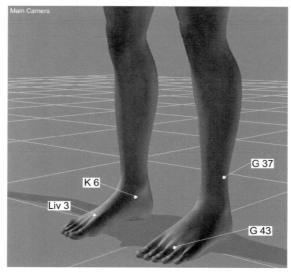

FIGURE 5.77 Recommended choice of APs for the treatment of blood supply–related ophthalmologic diseases/disorders.

such as attacks of acute angle–closure glaucoma, it is good to perform urgent treatments, with an immediate clinical improvement often observed. Otherwise, taking into account chronic nature of these diseases or disorders, treatments can be done three times per week over a prolonged period of time (at least 40 sessions; sometimes much more).

5.8.1.3.37 INSOMNIA/NEUROSIS

Insomnia can affect various aspects of human life and often leads to abuse of medication prescribed for the condition or even to drug addiction. It is often the result of stress and neurosis. Physical medicine, and especially combined acupuncture therapy, can be very helpful in the complex management of this problem; in general, relaxation and improved sleeping is actually a positive side effect of any acupuncture treatment, no matter what disease or disorder is treated!

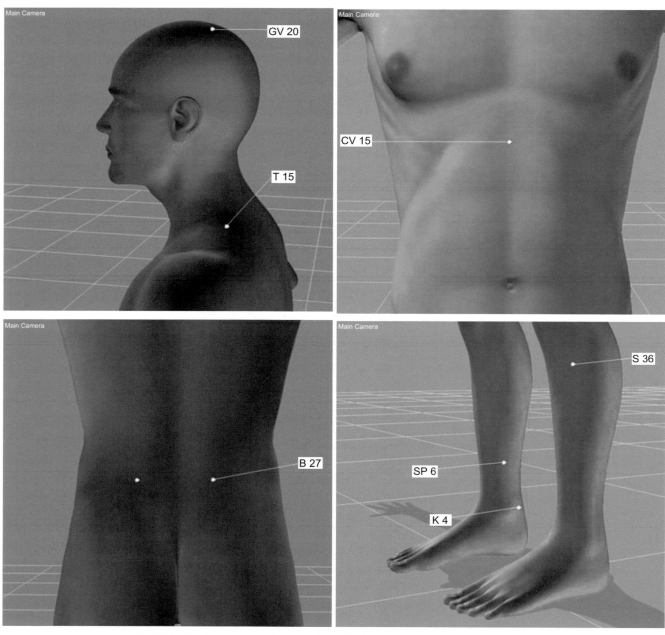

FIGURE 5.78 Recommended choice of APs for the treatment of neurosis and insomnia.

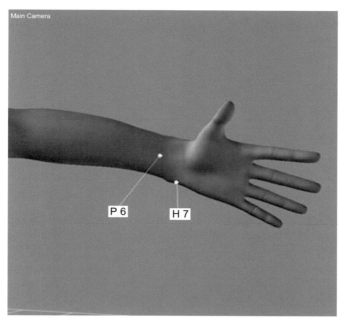

FIGURE 5.78 cont'd

However, there are specific APs that are most recommended for the management of neurosis and insomnia (Figure 5.78), subject to tenderness and "Acute"/"Subacute" readings on the OED device: T 15 and B 27 (because the patient undergoing a treatment lies on his or her back, both can be stimulated bilaterally with TENS), K 4, SP 6, S 36, H 7, P 6, CV 15, and GV 20. If available, electro-sleep therapy can be applied simultaneously (see Section 5.6.1.1), but in such a case the treatment duration has to be at least one hour instead of the standard stimulation time of 20 to 30 minutes. Treatments can be done three times per week, ideally if possible immediately before bed rest.

5.8.1.3.38 ANTINICOTINIC THERAPY

Properly performed electroacupuncture treatment can be, perhaps, the most effective antinicotinic therapy. In our clinical material of dozens of patients with nicotine addiction, we have achieved an approximate 85% success rate of permanently ending an intensive smoking habit.

The most recommended classical APs to be stimulated in the case of nicotine addiction, subject to tenderness and "Acute"/"Subacute" readings on the OED device, are as follows (Figure 5.79): T 15 (can be stimulated bilaterally with TENS), G 43, G 8, P 6, L 7, and GV 20. However, an additional (nonclassical) AP called Yintang should also be used; it is located on the Governing Vessel (GV), between the eyebrows. For "heavy" smokers, another two "extra" points can be added, which are located symmetrically on the outer sides of nostrils on the verge of nasal bones.

The recommended antinicotinic therapy also includes a very important auricular acupuncture: on the left ear auricle, OPAs corresponding to the bronchi and the stomach should be stimulated (see Figure 4.8); on the right one, OPAs related to the lungs (lower lobes) and the gallbladder should be stimulated. Auricular electroacupuncture can be done with small standard needles (3–4 cm long), which can be connected to the electrostimulator (see Section 5.8.1.2.3) along with all the other needles inserted into corporal APs.

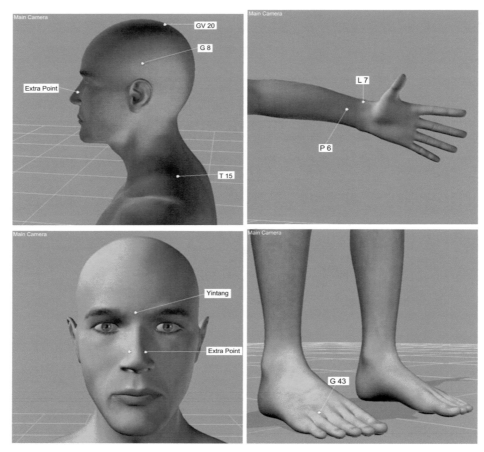

FIGURE 5.79 Recommended choice of APs for the anti nicotinic therapy.

This treatment seems to block the nicotinic hunger. Therefore, 10 minutes after the beginning of such a stimulation, the patient (still under treatment) should be asked to smoke one or two cigarettes; this will be like the first cigarette in his or her life, with intensive perspiration, dizziness, nausea, and sometimes vomiting and even fainting occurring before the smoking is finished. Then, following the famous example of Pavlov's dogs, the same procedure should be repeated three times per week, at least several times, until the patient develops a reflex permanently preventing him or her from attempting to smoke.

5.8.2 Auriculotherapy (Acupuncture of the Ear Auricles)

Auriculotherapy is the therapeutic stimulation of the OPAs located on the ear auricles. There is information in the oldest preserved Chinese manual of medicine, issued in the 4th century BC *Huangdi Nei Jing* (*Canon of Medicine*), that the ear is a meeting point of all the energetic channels of the human body. Therefore, ancient Far Eastern doctors tried to use the ear auricles for acupuncture purposes, but auricular acupuncture was never as popular as corporal acupuncture and finally was forgotten for many centuries. Certain European doctors also historically tried therapeutic stimulation of the ear auricles for specific conditions. In the 17th century, Portuguese doctor Zacutus Lusitanus apparently proposed auricular cauterisation for sciatica. In 1717, Italian doctor Valsalva described an auricular area to be punctured in the case of toothache. Cauterisation of the upper part of the ear auricle in the case of sciatica was recommended in the traditional medicine of the Middle East, France, and Italy.

More recently, the French doctor P.M.F. Nogier (42), working in North Africa, observed local healers treating domestic animals, including horses and camels, by cauterising particular zones on the animals' ears. They also used copper earrings, placed at specific auricular areas, for pain in humans. Inspired by this, Nogier examined ear auricles of his patients from 1951 to 1956 and concluded that a particular area would become tender when a related internal organ was diseased. In this way, he created the first maps of auricular OPAs and originated the concept of the auricular homunculus with the shape and position generally similar to the early fetus. He also employed this discovery for therapeutic purposes by inserting small acupuncture needles or applying laser radiation at these points. Auriculotherapy was then accepted by French and other European doctors interested in acupuncture, and finally also by Far Eastern classical acupuncturists. It is important that each auricular OPA corresponds to only one internal organ/body part, in contrast to corporal APs which are usually related to more than one internal organ/body part.

OED confirmed Dr Nogier's ideas in general, but because it is much more accurate than the basic palpation used by Nogier, it changed the locations of various OPAs (Figure 4.8). Because the distances between particular auricular OPAs can be very small, the OED examination should always be used for precise localization; only OPAs regarded as "Acute"/"Subacute" by the OED device should be stimulated.

Auricular OPAs can be stimulated with small acupuncture needles (standard length 3–4 cm) (Figure 5.80) in addition to the classical acupuncture treatment; these auricular needles can also be connected to the therapeutic electrostimulator (see Section 5.8.1.2.3). Special round microneedles or press needles with very short spikes can also be used; these can be placed

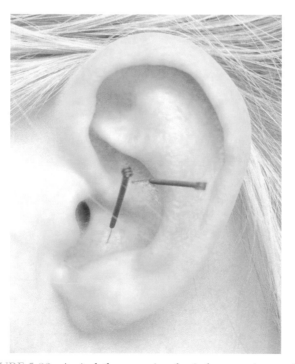

FIGURE 5.80 Auriculotherapy using classical acupuncture needles.

on the ear auricle and left there for seven to 10 days, after fixing them in place with a piece of self-adhesive tape (Figure 5.81). In such a case, the patient himself or herself should stimulate the auricular OPAs by applying finger pressure or soft massage to the microneedles, one by one, for several seconds: the procedure can be repeated three to four times per day. Instead of microneedles or press-needles, small magnets are sometimes fixed on ear auricles, but this kind of stimulation is very soft (see Section 5.7); only slightly stronger is laser stimulation (see Section 5.5.3).

Depending on the internal organ/body region treated, the respective auricular OPAs should be stimulated; for example, in the case of stomach pathology, it should be the OPA corresponding to the stomach that is stimulated, and in the case of knee problems, it should be the OPA related to the knee that is stimulated (see Figure 4.8). Organs/body parts located on the right side of the body, such as the right kidney or right elbow, should be treated by stimulation of the right auricular OPAs and vice versa; organs/body parts located on the left side of the body require left auricular OPA stimulation. Various auricular OPAs can be stimulated at the same time, as is recommended in case of, for example, antinicotinic therapy.

Auriculotherapy belongs to the group of very "soft" methods of reflexive physical medicine, and therefore there are practically no contraindications for its application. It is usually used as a supportive treatment to other more effective therapies, such as corporal acupuncture. Microneedles or press-needles placed at the OPAs related to the cortex can be tried for insomnia, anxiety, lack of concentration, and psychoses. Microneedles or press needles placed at the OPAs corresponding to the stomach (cardia on the left ear auricle or pylorus on the right one), gallbladder (right auricle), and pancreas (left auricle) can be helpful in the slimming therapies and should be manually stimulated for a view minutes whenever the patient feels hungry.

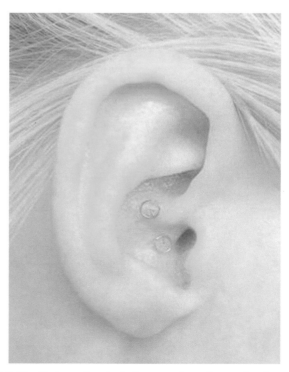

FIGURE 5.81 Prolonged auriculotherapy using microneedles/press needles.

5.8.3 Craniotherapy (Scalp Acupuncture)

Craniotherapy is the therapeutic stimulation of specific skin zones (not classical APs) located on the head. This method was invented in the 1970s by Chinese Neurologist Chiao Shun-fa, who initially experimented on himself using long needles inserted under the hairy cranial skin. He believed that stimulation of certain cranial skin zones will improve, in the reflexive way, the functioning of related cortical centers. Therefore, craniotherapy is aimed mainly at those diseases or disorders which are caused by cerebral pathologies.

Classical craniotherapy distinguishes several specific skin zones (see Figures 5.82–5.84):

1. Motoric zone and speech center I. Indications for stimulation: hemiparesis/hemiplegia, cerebral palsy (children), facial nerve palsy, motoric aphasia, speech disorders, hypersalivation.
2. Perception zone. Indications for stimulation: perception disorders of central origin, trigeminal neuralgia, toothache.
3. Anticonvulsant zone. Indications for stimulation: Parkinsonism, chorea.
4. Vasomotor zone. Indications for stimulation: edema of vasomotor origin, such as after stroke.
5. Acoustic and vertigo zone. Indications for stimulation: hearing loss, tinnitus, dizziness of labyrinthine origin, Ménière's syndrome.
6. Speech zone II. Indication for stimulation: complete sensory aphasia.
7. Speech zone III. Indication for stimulation: sensory aphasia.
8. Apraxia zone. Indication for stimulation: apraxia.
9. Gait coordination zone. Indications for stimulation: imbalance, walking difficulties of central origin, painful legs.
10. Visual zone. Indication for stimulation: sight disturbances of cortical origin.

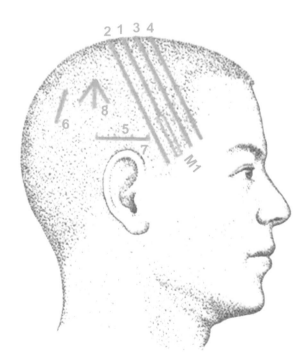

FIGURE 5.82 Classical craniotherapeutic zones on the lateral side of the head. 1–motoric zone, MI–speech centre I, 2–perception zone, 3–anticonvulsant zone, 4–vasomotoric zone, 5–acoustic and vertigo zone, 6–speech zone II, 7–speech zone III, 8–apraxia zone.

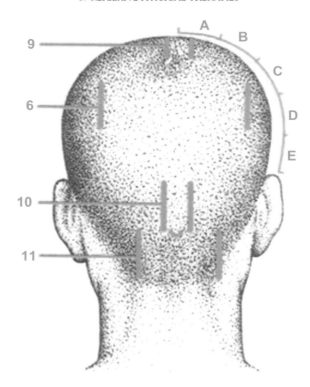

FIGURE 5.83 Classical craniotherapeutic zones on the posterior side of the head. 6–speech zone II, 9–gait coordination zone, 10–visual zone, 11–cerebellar balance zone. All of the main craniotherapeutic zones (1–4, see: Figure 5.82) are divided for five subcones related to the particular body regions.

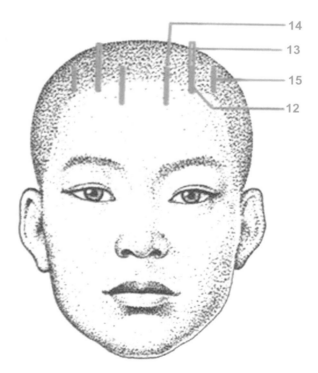

FIGURE 5.84 Classical craniotherapeutic zones on the forehead and anterior part of the fornix. 12–gastric zone, 13–liver and gallbladder zone, 14–chest zone, 15–genital zone.

11. Cerebellar balance zone. Indication for stimulation: balance disturbances of cerebellar origin.
12. Gastric zone. Indications for stimulation: peristaltic disturbances and other gastric malfunctions.
13. Liver and gallbladder zone. Indications for stimulation: Liver and gallbladder diseases.
14. Chest zone. Indications for stimulation: bronchial asthma, ischaemic heart disease.
15. Genital zone. Indications for stimulation: genital functional disturbances.

During a craniotherapeutic session, depending on the location of the respective skin zone, the patient can be in a sitting position (increased risk of fainting) or can lay on his or her abdomen or back. If the treated pathology is located on the right side of the body, then the respective skin zone should be stimulated on the left side of the head and vice-versa: only tender zones—OED reading of "Acute"/"Subacute"—should be stimulated. Depending on the length of the particular zone, longer acupuncture needles (6.5–10 cm) should be inserted subcutaneously (between the skin and the skull), one by one, along the zone. The needles can be then stimulated electrically or thermally, in the same way as in the case of corporal acupuncture (see Section 5.8.1.2.3). Craniotherapy can be used simultaneously with other methods of reflexive physical medicine.

Contraindications for craniotherapy are the same as for acupuncture in general (see Section 5.8.1.2.4). Stimulation of genital zones should be avoided during pregnancy due to a potential risk of premature delivery.

5.8.4 Reflexive Massage

Not everybody is courageous enough to undergo therapeutic needling. In such a case, especially when the treated pathology is not a very serious one, reflexive massage can be tried. This kind of massage is aimed at the manual stimulation of the appropriate trigger/pressure points (Ashi points), and it has a long history; there are indications that, under the name of Tien-An therapy, it was developed in ancient China in 4000 years BC; in Japan, it was known as Shiatsu and existed from the 6th century. Reflexive point massage is often called acupressure in contemporary times, from Latin *acus*—needle—and *presso*—pressure, despite the fact that no needle is used. From a linguistic point of view, another contemporary term is equally strange: Pressopuncture, which is used despite the fact that no puncture takes place.

Although the selection of the most suitable skin areas is similar to that of the acupuncture system, only the most tender spots/zones are to be stimulated with relatively strong pressure, circular kneading, and even pulling, using one, two, or three fingers or even the whole hand. Each skin spot/zone should be stimulated, one by one, for at least several seconds; the stimulation of most important spots/zones can then be repeated. The usual duration of the complete therapeutic session is 15 to 45 minutes. In more acute cases, treatments can be done two to three times per day; in more chronic cases, treatment can be performed on a daily basis. Rubbing with therapeutic oils or creams, which stimulate skin nervous receptors in a chemical way, will enhance the therapeutic effect.

For basic reflexive massage treatments, the skin areas indicated in Figure 5.9 can be used; this can also be a form of autotherapy, especially in the case of pain syndromes or sports medicine. There are no direct contraindications to reflexive massage therapy, but it should not be applied to damaged skin or varicose veins. Intensive reflexive massage should not be used on the lower back, abdomen, and legs of pregnant women due to a potential risk of induction of premature delivery.

5.8.4.1 Reflexology (Reflexive Massage of Feet/Hands)

Linguistically, the term "reflexology" appears to be very unfortunate, because the component *logy* (from the Greek *logia*) suggests science, doctrine, or theory. Instead, in English-speaking countries, reflexology became a popular synonym for the reflexive massage of feet or hands. To make it even stranger, reflexologists usually use the pseudoscientific term "receptor" when referring to the OPAs located on these body parts.

Reflexive massage of feet or hands is based on the belief that the OPAs related to all the internal organs/body parts are located on the surface of the feet and hands (Figure 5.85), as in the case of ear auricles. When corresponding organs are diseased, the respective OPAs become tender; there have been attempts to use this phenomenon for diagnostic purposes. However, strict clinical trials are still required; the OED would perhaps be very useful, but the usually thick and dry epidermis of plantar areas, which significantly increases the skin electrical resistance, might create some measuring difficulties. The treatment involves manual stimulation of all the tender zones, in exactly the same way which was described in the previous section.

Reflexive massage of feet and hands belongs to the "soft" methods of physical medicine; therefore, it is mainly used for recreational purposes. It should not be applied to damaged skin.

5.8.4.2 Tsubo Therapy

This specific form of reflexive massage originating from Japan is based on the acupuncture system, but the needles are replaced by small metal spheres fixed to the respective APs by means of small pieces of adhesive tape and left there until the whole course of treatments is completed. The therapist or patient himself or herself stimulates the APs by pressing the spheres, one by one, for several seconds; the procedure can be repeated three to four times per day. Tsubo therapy can include both corporal APs and auricural OPAs, but the number of stimulated skin spots should be limited for practical reasons.

Pieces of adhesive tape can be replaced as necessary, usually every seven to 10 days.

Tsubo therapy belongs to the very "soft" methods of reflexive physical medicine, and there are practically no contraindications to its application. However, it should not be applied to damaged skin.

5.8.5 Cupping

Cupping belongs to the oldest therapeutic methods and was known in all ancient civilizations. In the modern world, it is still among the most popular home treatments. The suction cups were originally made from animal horns or bamboo, and later were made of glass; this allowed observation of the skin undergoing treatment. Reflexive physical medicine utilizes only the so-called dry cupping; there are no skin cuts under the cup, unlike so-called wet cupping, which is used mainly to suck out something such as snake or insect toxins.

Cupping therapy mechanically stimulates skin nervous receptors in a way similar to reflexive massage; instead of pressure, however, it uses a vacuum effect. This can be achieved either by heating the air inside the cup with a flame or using a vacuum pump connected to the cup via flexible pipes with a manometer to control the pressure.

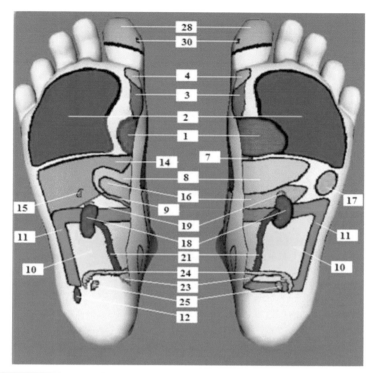
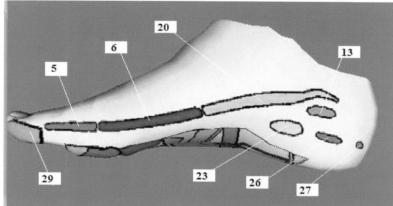
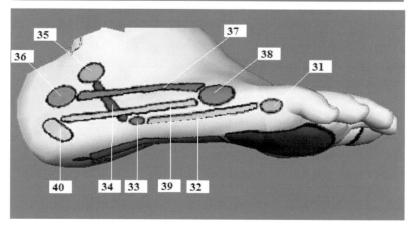

FIGURE 5.85 Proposed map of the OPAs located on feet (locations not proven clinically). Pedal OPAs located on the left foot correspond to left-sided internal organs/body parts, whereas pedal OPAs located on the right foot are related to the right-sided organs/body parts. **Chest and neck:** 1–heart, 2–lung, 3–oesophagus, 4–thyroid gland, 5–cervical spine, 6–thoracic spine. **Abdomen:** 7–cardia, 8–stomach, 9–duodenum, 10–small intestine, 11–colon, 12–appendix, 13–rectum, 14–liver, 15–gall-bladder, 16–pancreas, 17–spleen, 18–kidney, 19–adrenal glands, 20–lumbo-sacral spine. **Pelvis:** 21–ureter, 22–urinary bladder, 23–uterus, 24–fallopian tube, 25–ovary, 26–vagina, 27–testis. **Brain:** 28–cortex, 29–midbrain, 30–pituitary. **Upper limbs:** 31–shoulder, 32–humerus, 33–elbow, 34–forearm, 35–hand. **Lower limbs:** 36–hip, 37–femur, 38–knee, 39–lower leg, 40–foot.

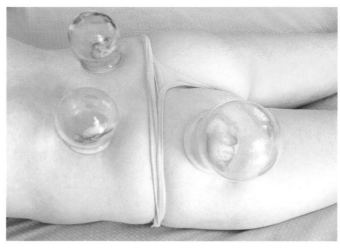

FIGURE 5.86 Hot cupping therapy using traditional glass cups.

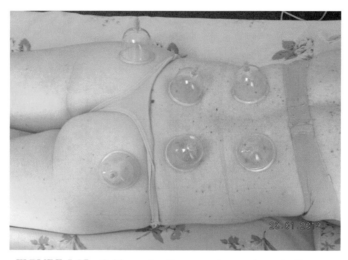

FIGURE 5.87 Cold cupping therapy using modern plastic cups.

In the case of hot cupping, a burning, spirit-imbibed swab should be inserted inside the cup for three seconds, and then the cup should be quickly placed on the chosen skin area (Figure 5.86); this kind of treatment also includes a thermal stimulation, to some extent. In the case of cold cupping, the pressure inside the cup should be established at about 0.2 atmosphere (Figure 5.87). Four to 10 cups are usually used simultaneously. Skin areas chosen for cupping therapy should be even; they are usually above large muscles, in order to insure good fixation of the cup. Only tender skin areas, with OED readings of "Acute"/"Subacute" should be stimulated. Prior to treatment, the skin should be cleaned with water, soap, and high-percentage alcohol and covered with petrolatum. The typical duration of the cupping treatment varies from 15 to 25 minutes. Treatments can be done on a daily basis for acute problems and every other day for more chronic conditions.

Cupping is among the more effective physical therapies within its range of indications, which traditionally include pneumonia, acute bronchitis, bronchial asthma, back pains, and neuralgias. The cups can be placed on the respective skin areas indicated in Figure 5.9.

CHAPTER 6

Final Considerations

The research described in this book provided a universal and scientifically acceptable explanation of the neurophysiological foundation of all methods of reflexive physical medicine. It also spread new light on pain mechanisms and our understanding of the functioning of the nervous system, especially the sensory nervous system.

The sensory nervous system—the body's primary information network—is the best first-line diagnostic system known. It can detect any damage done to the body, from both outside and inside, at an early stage, and it can also estimate the seriousness of the problem; it sends a weak nervous signal to the decision-making centers of the CNS in the case of, for example, chronic asymptomatic gastritis, but it sends a much stronger one in the case of, for example, a perforated peptic ulcer. The activation of the body's first-line potent self-defense mechanisms depends on this information. Controlling the human body's information system therefore creates entirely new opportunities for both diagnostics and therapy.

For centuries, due to limited technical means, doctors had no option but to make use of both diagnostics and therapy "from outside." However, the OED opened the door to priceless diagnostic information circulating in the sensory nervous system, and reflexive therapies can "force" the powerful self-defense mechanisms of the body to act in a desired way. It has to be emphasized that reflexive physical medicine is not only natural, it is also safe; it might not help when used beyond the scope of its indications, but it cannot aggravate the problem. Therefore, physical medicine should no longer be treated as the "Cinderella" of contemporary medicine, but rather as the medicine of the future (with ancient roots)! Hopefully, a new generation of researchers will take physical medicine even further through this newly opened door.

Unfortunately, in many countries physical medicine is still not recognized as an official medical specialty. Certain methods of physical medicine are even labeled as "alternative"/"integrative," "complementary," "regulatory," or "unconventional." Medicine is all one discipline; it can be bad, such as "failed spinal surgery," or it can be good, such as when allergic rhinitis is cured by a homeopathic remedy, but it is always the same medicine! Therefore, there is no need for the terms "alternative" or "integrative." Physical medicine cannot be regarded as "complementary" because in many diseases or disorders it should be the first-line therapy, with medication and other therapies being only a supportive treatment. When it comes to the term "regulatory," all effective therapies, including pharmacotherapy, can in fact be labeled this way. The convergence modulation theory provides a clear, universal, scientifically acceptable explanation of the mode of action of all methods of reflexive physical medicine; therefore, they should no longer be labeled "unconventional." It may be practical to combine physical medicine and medical rehabilitation into one specialization called, for example, "physical medicine and rehabilitation," but it has to be clearly understood that these are two separate medical specialties with different physiological foundations. In fact, such a fusion would be similar to "internal medicine and neurology" or "surgery and gynecology."

Academic societies now must take seriously the great scientific and practical potential of reflexive physical medicine, and medical authorities in countries with no official specialization in physical medicine should consider its introduction. On the other hand, schools of traditional Chinese medicine can no longer pretend that nothing has changed in medicine during the last three thousand years. Unscrupulous or naïve manufacturers or distributors of many supposed physical medicine products have to be much more responsible with their offerings to the public; the production and distribution, sometimes at high cost, of nonsense diagnostic and therapeutic devices that are clinically unproven is unethical and against the rules of good business!

References

[1] Bischko J. An Introduction to Acupuncture. Heidelberg: Haug Verlag; 1978.
[2] Bong-Han K. On the Kyungrak System. DPRK: Foreign Languages Publishing House; 1964.
[3] Brain SD, Newbold P, Kajekar R. Modulation of the release and activity of neuropeptides in the microcirculation. Can J Physiol Pharmacol 1995;73:995.
[4] Bossy J. Morphological data concerning the acupuncture points and channel network. Acup Electro-Ther Res Int J 1984;9:79.
[5] Bratu J, Prodescu V, Georgescu A. Kortikale Behandlung durch Akupunktur. Dtsch Z Akup 1954;3:113.
[6] Buettner W. Literaturbetrachtungen und abschliessende Bemerkungen zum Elektroherdtest-Problem. Dtsch Stom 1965;15(No. 11):847.
[7] Cheng RSS, Pomeranz B. Electroacupuncture analgesia could be mediated by at least 2 pain relief mechanisms—Endorphin and Non-Endorphin systems. Life Sci 1979;25(No. 23):1957.
[8] Chernomordik LV. Breakdown of lipid bilayer membranes in an electric field. Biochim et Biophys Acta 1983;739:203.
[9] Chizmadzhev Y, Zamitsin V, Weaver JC, Potts R. Mechanism of electroinduced ionic species transport through a multilamellar lipid system. Biophys J 1995;68:749.
[10] Ciszek M, Szopinski J, Skrzypulec V. Investigations of morphological structure of acupuncture points and meridians. J Tradit Chin Med 1985;5:289.
[11] Clement-Jones V, McLoughlin L, Lowry PJ, Besser GM, Reess LH. Acupuncture in heroin addicts: Changes in Met-Enkephaline and Beta-Endorphin in blood and cerebrospinal fluid. Lancet 1979;25:380.
[12] de Wet EH, Oosthuizen JMC, Odendaal SL, Shipton EA. Neurochemical mechanisms that may underlay the clinical efficiency of AP simulation therapy in chronic pain management. South Afr J Anaest Analg 1999;5:33.
[13] Dominikowska A. Estimation of the ETS test usefulness in the diagnostics of children sinusitis (Polish). Ph.D. thesis. Warsaw Medical Academy; 1972.
[14] Domzal T. Pain (Polish). Warsaw: WP; 1980.
[15] Durinian R. Physiological basis of auricular reflex therapy. Erevan: Ayastan Publishing House; 1983.
[16] Durinian R. Reflexive changes in spontaneous and evoked activity of neurons of the feline thalamic parafascicular complex due to electroacupuncture stimulation (Russian). Biul Eksp Biol Med 1983;96:3.
[17] Edmond S. The Listen System. Brampton, Ontario: The Printing Team; 1997.
[18] Fletcher RH, Fletcher SW, Wagner EH. Clinical Epidemiology, The Essentials. Baltimore: Williams & Wilkins; 1991.
[19] Franek A, Franek E, Polak A. Modern electrotherapy (Polish). Katowice: Dzial Wydawnictw SAM; 2001.
[20] Garnuszewski Z. Acupuncture in Contemporary Medicine (Polish). Warsaw: Amber; 1997.
[21] Gavrylov A, Saharovskaya L, Sinelstchykov A, Sokolov S, Gunik A. Contemporary problems in reflexive diagnostics and therapy (Russian); 1984. Paper presented at the 1st Applied Science Conference. Rostov-on-Don, 14–16 June.
[22] Glaser R, Leikin S, Chernomordik L, Pastushenko V, Sorkirko A. Reversible electrical breakdown of lipid bilayers: formation and evolution of pores. Biochim et Biophys Acta 1988;940:275.
[23] Glaser-Tuerk M. Specielle lokale Teste zum Nachweis aktiver Herde im Kieferbereich. Oester Z Stom 1968;65:222.
[24] Grimnes S. Skin impedance and Electro-Osmosis in the human epidermis. Med Biol Eng Comput 1983;21:739.
[25] Hyodo MD. Ryodoraku Treatment and Objective Approach to Acupuncture. Osaka; 1975.
[26] Jonderko G, Szopinski J, Galaszek M, Galaszek Z. Einfluss von gleichzeitiger Warme- und Kaltetherapie auf die Schmerzschwelle und auf die subjektive Schmerzbeurteilung bei Patienten mit chronischer Polyarthritis. Z Phys Med Baln Med Klim 1988;17:369.
[27] Jonderko G, Szopinski J, Galaszek M, Galaszek Z. The influence of heating and cooling procedures on pain perception of patients with rheumatoid arthritis (Polish). Anest Inten Ter 1990;22:41.
[28] Jonderko G, Szopinski J, Miarka J. Reflexive therapy in diabetic polyneuropathy (Polish). Anest Inten Ter 1982;14:339.
[29] Kellner G. Uber ein vaskularisiertes Nervenendkorperchen vom Typ der Krauseschen Endorgane. Zschr Mikr Anat Forsch 1966;1:130.

[30] Kenyon JN. Diagnostic techniques using bioenergetic recording methods: The science of bioenergetic regulatory medicine. Am J Acup 1986;14:5.
[31] Ketterl W. Untersuchungen uber den Elektro-Herd-Test. Dtsch Zahnarztl Z 1962;17(No. 7):503.
[32] Kleiner M. Biochemistry. ed. 6 The C V Mosby Company; 1962.
[33] Koehnlechner M. Handbuch der Naturheilkunde. Munich: Kindler Verlag; 1975.
[34] Koenig G, Wancura I. Einfuehrung in die chinesische Ohrakupunktur. Heidelberg; 1978.
[35] Lange G. Akupunktur der Ohrmuschel. Diagnostik und Therapie. Schorndorf: WBV Biologisch-Medizinische Verlags GmbH & Co KG; 1985.
[36] Lochner G. The voltage-current characteristic of the human skin. Master Dissertation. University of Pretoria; 2003.
[37] Lotti L, Hautmann G, Panconesi E. Neuropeptides in skin. J Am Acad Dermatol 1995;33:482.
[38] Melzack R, Wall PD. Pain mechanisms: a new theory. Science 1965;150:45.
[39] Mika T, Kasprzak W. Physical therapy (Polish). Warsaw: Wydawnictwo Lekarskie PZWL; 2001.
[40] Misery L. Neuro-immuno-cutaneous system (NICS). Pathol Biol 1996;44:867.
[41] Niboyet J. Complements d'Acupuncture; 1955. Paris.
[42] Nogier PMF. Handbook of Auriculotherapy. Moulins-les-Metz: Maisonneuve; 1981.
[43] Panescu D, Webster JG, Stratbucker RA. A nonlinear electrical-thermal model of the skin. IEEE Trans Biomed Eng 1994;41:672.
[44] Pauser G, Gilly H. Neurophysiologische und neurohumorale Mechanismen der Akupunkturanalgesie. In: Bischko J, editor. Handbuch der Akupunktur und Aurikulotherapie, Part 25.2.0. Heidelberg: Haug Verlag; 1985.
[45] Pliquett UF, Gusbeth CA. Perturbation of human skin due to application of high voltage. Bioelectrochemistry & Bioenergetics 2000;51:41.
[46] Pomeranz B. Acupuncture reduces electrophysiological and behavioral responses to noxious stimuli: Pituitary is implicated. Exp Neurol 1977;54(No. 1):172.
[47] Pomeranz B. Electroacupuncture hypalgesia is mediated by afferent nerve impulses: An electrophysiological study in mice. Exp Neurol 1979;66(No. 2):398.
[48] Portnov F. Electropuncture Reflexotherapy (Russian). Riga: Zinatne Publishing House; 1982.
[49] Potts RO, Guy RH. Routes of ionic permeability through the mammalian skin. Solid State Ionics 1992;53–56:166.
[50] Reshetnyak W. Changes in bioelectrical activity in the orbito-frontal and somatosensory regions of the cortex due to electroacupunture stimulation (Russian). Biul Eksp Biol Med 1982;93:5.
[51] Reshetnyak W. Reflexive changes of the functional activity of the central cortex due to electroacupuncture stimulation (Russian). Biul EkspBiol Med 1983;96:14.
[52] Riederer P, Tenk H, Werner H. Biochemische Aspekte der Akupunktur. Dtsch Z Akup 1978;2:59.
[53] Rosenblatt S. The electrodermal characteristics of acupuncture points. Am J Acup 1982;10:131.
[54] Schmidt H. Akupunkturtherapie nach der chinesischen Typenlehre. Stuttgart: Hippokrates Verlag GmbH; 1978.
[55] Sierak T, Szopinski J. Universal device for organ electrodermal diagnostics and electrotherapy (Polish). Probl Tech Med 1987;18:255. 1987.
[56] Sierak T, Szopinski J. Universelles elektronisches Gerat fur die automatische Elektropunkturdiagnostik und Elektroreflextherapie. Dtsch Z Akup 1988;5:112.
[57] Sieron A, Cieslar G, Krawczyk-Krupka A, Biniszkiewicz T, Bilska-Urban A, Adamek M. Utilization of the magnetic fields in medicine. Theoretical foundations, biological effects and clinical applications. (Polish). In: Sieron A, editor. Bielsko-Biala. 2nd ed. Alpha-Medica Press; 2002.
[58] Sjolund B, Eriksson M. Electro-acupuncture and endogenous morphins. Lancet 1976;2:7994.
[59] Sjolund B, Terenius L, Eriksson M. Increased cerebrospinal fluid levels of endorphins after electro-acupuncture. Acta Physiol Scand 1977;100(No. 3):382.
[60] Szopinski J. Application of the organ electrodermal diagnostics own method in chosen internal diseases (Polish). Ph.D. thesis. Katowice: Silesian Medical Academy; 1985.
[61] Szopinski J. The use of bioelectrical properties of skin for organ diagnostics (Polish). Wiad Lek 1989;42:697.
[62] Szopinski J, Lukasiewicz S, Lochner G, Nasiek D, Krupa-Jezierska B, Warakomski P, Kielan K. Influence of general anaesthesia and surgical intervention on the electrical parameters of auricular organ projection areas. Med Acup 2003;14(No. 2):40.
[63] Szopinski J, Lochner G, Szkliniarz J, Karcz-Socha I, Kasprzyk-Minkner A, Kielan K, Krupa-Jezierska B, Nasiek D, Warakomski P. Localization of auricular projection areas of the stomach and duodenum and their use in the monitoring of ulcer disease. Med Acup 2003;15(No. 1):31.

[64] Szopinski J, Lochner G, Macura T, Karcz-Socha I, Kasprzyk-Minkner A, Kielan K, Krupa-Jezierska B, Nasiek D, Warakomski P. Localization of auricular projection area of the liver and its use in the monitoring of viral hepatitis. J Trad Chin Med 2006;27(No. 3).
[65] Szopinski J, Lochner G, Pantanowitz D. Influence of organ pathology on the electrical parameters of organ projection areas of the skin. J Trad Chin Med 2006;26(No. 3):218.
[66] Szopinski J, Lochner G, Szopinska H. The effectiveness of analgesic electrotherapy in the control of pain associated with diabetic neuropathy. South Afr J Anaest Analg 2002;8(No. 4):12.
[67] Szopinski J, Pantanowitz D, Jaros GG. Diagnostic accuracy of organ electrodermal diagnostics. Pilot study SAMJ 1998;88:146.
[68] Szopinski J, Pantanowitz D, Lochner G. Estimation of the diagnostic accuracy of organ electrodermal diagnostics. SAMJ 2004;94(No. 7):547.
[69] Szopinski J, Pastor A. Patent No 2001/0970; 2001. South Africa.
[70] Szopinski J, Sierak T, Gabryel A. Die Erforschung der Bioenergetischen Eigenschaften der Akupunkturmeridiane. Dtsch Z Akup 1988;2:31.
[71] Szopinski J, Sierak T, Lochner G. Neurophysiological foundations of organ electrodermal diagnostics, acupuncture, TENS and other reflexive therapies. South Afr J Anaest Analg 2004;10(No. 3):21.
[72] Szopinski J, Sierak T, Kaniewski M, Niezbecki A. Der Einfluss ausgewaehlter Krankheiten auf die Bioelektrischen Eigenschaften der Akupunkturpunkte. Dtsch Z Akup 1988;3:51.
[73] Szopinski J, Sierak T, Kaniewski M, Niezbecki A. Investigations into the effects of particular electrical current parameters on acupuncture points in order to determine the optimal parameters for reflexive electrotherapy (Polish). Akup Pol 1983;1:17.
[74] Szopinski J, Sierak T, Kaniewski M, Niezbecki A. Untersuchung der bioelektrischen Erscheinungen, die wahrend der reflextherapeutischen Behandlung auftreten. Dtsch Zschr Akup 1988;3:56.
[75] Szopinski J, Sierak T, Niezbecki A, Kaniewski M. Influence of selected internal diseases on electrical parameters of acupuncture points (Polish). In: Sapinski W, editor. Diary of 1st National Conference on Acupuncture, 23–24 September 1982, Warsaw. Kalisz: WSZ; 1982.
[76] Szopinski J, Sierak T, Szopinska H, Ciszek M. Bioelektrische und bioenergetische Eigenschaften und morphologische Strukturen der Akupunkturpunkte und Akupunkturmeridiane. In: Bischko J, editor. Handbuch der Akupunktur und Aurikulotherapie. Part 25.2.0. Heidelberg: Haug Verlag; 1985.
[77] Stux G, Stiller N, Pothmann R, Jayasuriya A. Lehrbuch der klinischen Akupunktur. Berlin; Heidelberg; New York: Springer-Verlag; 1981.
[78] Tabeyeva O. Manual of Acupuncture Reflexotherapy (Russian). Moscow: Meditsina; 1980.
[79] Tashkin DP, et al. Comparison of real and simulated acupuncture and Isoproterenol in Metacholine-induced Asthma. Ann Allergy 1977;39(No. 6):379.
[80] Toda K. Effects of electroacupuncture on thalamic evoked responses. Exp Neurol 1979;66(No. 2):419.
[81] Toman K. Sensitivity, specificity and predictive value of diagnostic tests. Bull Int Union Tuberc 1981;56:15. 1981.
[82] Traczyk W, Trzebski A. Human Physiology with Elements of Clinical Physiology (Polish). Warsaw: PZWL; 1980.
[83] Umlauf R. Zu den Grundmechanismen der Akupunkturwirkung und Moglichkeiten ihrer Beeinflussung. Heidelberg: Haug Verlag; 1988.
[84] Umlauf R. Zu den wissenschaftlichen Grundlagen der Aurikulotherpie. Dtsch Zschr Akup 1988;3:59.
[85] Vandan Y, Zaltsmanie V. Morphological characteristics of the biologically active points (Russian). In: Problems of Clinical Biophysics. Riga; 1977.
[86] Vander A, Sherman A, Luciano D. Human Physiology. ed. 7 New York: WCB McGraw-Hill; 1998.
[87] Vandoorne K, Addadi Y, Neeman M. Visualizing vascular permeability and lymphatic drainage using labeled serum albumin. Angiogenesis 2010;13:75.
[88] Voll R. Topographische Lage der Messpunkte der Elektroakupunktur. Uelzen: Dr. Blume & Co. (ML Publishers); 1968.
[89] Wall PD. The gate control theory of pain mechanism, a reexamination and re-statement. Brain 1978;101:1.
[90] Weaver JC, Vaughan TE, Chizmadzhev Y. Theory of electrical creation of aqueous pathways across skin barriers. Adv Drug Deliv Rev 1999;35:21.
[91] Welhover E, Romashov F. Psychological Self-Control (Russian). Alma-Ata 1974.
[92] White M. Principles of Biochemistry. 5th ed. New York: McGraw-Hill; 1973.
[93] Wu CC, Hsu CJ. Neurogenic regulation of lipid metabolism in rabbits: A mechanism for the cholesterol-lowering effect of acupuncture. Atherosclerosis 1979;33(No. 2):153.

[94] Yamamoto T, Yamamoto Y. Electrical properties of the epidermal stratum corneum. Med Biol Eng 1976;14:151.
[95] Yamamoto T, Yamamoto Y. Non linear electrical properties of the skin in the low frequency range. Med Biol Eng Comput 1981;19:302.
[96] Zimmerman M. Physiologische Mechanismen von Schmerz und Schmerztherapie. Triangel 1981;20:7.

Index

Note: Page numbers followed by "f" indicate figures; "t", tables; "b", boxes.

A
Abdomen, acute, 75
Achalasia. *See* Esophagospasm
Acupressure. *See* Pressopuncture
Acupuncture, 109
  acoustic nerve damage, 199, 200f
  ancient Chinese medicine, 110
  ankle pain, 193–194, 194f
  antinicotinic therapy, 213–214, 214f
  auriculotherapy, 214–216, 215f–216f
  back pain with spinal fusion, 172
  bronchial asthma, 201–203, 202f
  C3 neuralgia, 179–181, 180f
  Chinese pathology, 110
  clinical effectiveness, 109
  combined acupuncture treatment, 165–166
  contraindications, 166
  cosmic principles, 110
  craniotherapy, 217–219
    forehead and anterior part of fornix, 218f
    lateral side of head, 217f
    posterior side of head, 218f
  cun, 111, 111f
  cupping, 220–222
    using glass cups, 222f
    using plastic cups, 222f
  CV, 138f–139f, 160b–161b
  delayed delivery, 210
    induction of labor uteral contraction, 210f
  dermatomes, 163f–164f
  diabetic neuropathy, 194–196, 195f
  elbow pain, 189–190, 189f
  electroacupuncture, 166f
  esophagospasm, 203, 203f
  fibromyalgia, 186–187
  gallbladder meridian, 132f–133f, 155b–157b
  globus hystericus, 204, 204f
  GV, 136f–137f, 159b–160b
  headache, 182–186
    cervicogenic treatment, 185f
  heart meridian, 120f–121f, 146b
  hip pain, 191, 192f
  hyperemesis gravidarum, 209, 209f
  insomnia, 212–213, 212f–213f
  intercostal neuralgia, 175f, 176
    exemplary location of PPs/TPs, 177f
  kidney meridian, 126f–127f, 152b–153b
  knee pain, 191–192, 193f
  L1 neuralgia, 172–173, 174f
  L2 neuralgia, 172, 173f
  LI meridian, 114f–115f, 140b–141b
  liver meridian, 134f–135f, 158b
  lung meridian, 112f–113f, 140b
  Meniere's disease, 200, 201f
  myofascial pain, 187
  needles, 162–165, 164f
  neurogenic bladder, 207–208, 207f
  nocturnal enuresis, 208–209, 208f
  optic nerve atrophy, 210–212, 211f
  pericardium meridian, 128f–129f, 153b
  phantom pain, 187
  procedures, 162–165
  PSSO, 167–168
  Raynaud's syndrome, 196, 197f
  reflexive massage, 219
    map of OPAs, 221f
    reflexology, 220
    Tsubo therapy, 220
  renal colic, 205–206, 206f
  rhinitis, 196–197, 198f
  selection, 162
  shoulder pain, 188–189, 188f
  SI meridian, 122f–123f, 146b–147b
  sinusitis, 198–199, 199f
  SP meridian, 118f–119f, 144b–145b
  stomach meridian, 116f–117f, 142b–144b
  T11–12 neuralgia, 175–176, 175f
  topography of meridians, 111
  trigeminal neuralgia, 181, 183f
    mandibular division treatment, 182f
    maxillary division treatment, 182f
    ophthalmic division treatment, 183f
  triple warmer meridian, 130f–131f, 154b–155b
  ulcerative colitis, 204–205, 205f
  upper back pain, 177–178, 178f
    brachialgia treatment, 179f
  urinary bladder meridian, 124f–125f, 148b–151b
  vital energy, 110–111
  wrist pain, 190–191, 190f
Acupuncture meridian
  bioelectrical properties, 24, 26
    classical acupuncture rules, 25
    clinical diagnoses, 24
    electrical impedance, 24–25, 25f
    electrical resistance, 25, 25f
    insertion of needles, 26
    internal organ pathology, 24
  histomorphological structure, 26–27, 27f
    APs' localization, 26
    Kyungrak system, 26
    skin samples, 27
  radioisotopic investigation, 12–13, 13f, 28
Acupuncture point (AP), 11

Adenosine triphosphate (ATP), 75–76
Aesthetic mode, 90
Albumin molecules, 43f
Alternative medicine, 225
Ankle pain, 193–194, 194f
Antinicotinic therapy, 213–214, 214f
AP. *See* Acupuncture point
Atonic bladder. *See* Neurogenic bladder
ATP. *See* Adenosine triphosphate
Auricular projection area
  esophagus, 65f
  evidence-based map, 57f
  gallbladder, 62f
  liver and monitoring of viral hepatitis, 60–61
  rectification ratios, 59f–60f
  stomach and duodenum, 57–59, 58f
Auriculotherapy, 214–216, 215f–216f

B
Back pain, 172. *See also* Low back pain
  acupuncture treatment, 172f
  in patients with spinal fusion, 172
Balneotherapy, 6
Benign prostate hypertrophy (BPH), 66–70
Bernard's current, 88–89
Bio-energetic therapies, 33
Blocked electropore, 44f
Brachialgia treatment, 179f
Bronchial asthma, 82, 201–203, 202f
Bronchospasm, 201–202

C
C3 neuralgia, 179–181, 180f
Cable electrodes, 101–102
Carpal tunnel syndrome, 190–191, 190f
Central nervous system (CNS), 35
Cervicogenic treatment, 185f
Chinese pathology, 110
Chronic condition therapy, 90
Chronic fatigue syndrome, 186
Chronic obstructive pulmonary disease (COPD), 104
Chronic sinusitis, 82
Climatotherapy, 6
Cluster headache, 184
CNS. *See* Central nervous system
Cohesion, 86
Collimation, 86
Complementary medicine, 225
Computed tomography scan (CT scan), 64
Conception vessel (CV), 138f–139f, 160b–161b

231

Convergence modulation theory, 40, 40f
  albumin molecules, 43f
  biological action, 40
  dermal nerve receptors, 41
  electropore, 43f
  nociceptive nervous receptors, 40–41
  outermost layers of skin, 42f
  reflexive therapies, 41
COPD. *See* Chronic obstructive pulmonary disease
Corona discharge photography. *See* Kirlian photography
Craniotherapy, 217–219
  forehead and anterior part of the fornix, 218f
  lateral side of head, 217f
  posterior side of head, 218f
Cryotherapy, 78
CT scan. *See* Computed tomography scan
Cun, 111, 111f
Cupping, 220–222
  using glass cups, 222f
  using plastic cups, 222f
CV. *See* Conception vessel

D
D&C. *See* Dilatation and curettage
D'Arsonval's currents, 100
Dark discharges, 100
Delayed delivery, 210
  induction of labor uteral contraction, 210f
Dermal nerve receptor, 41
Dermal nervous receptor, 36
Diabetic neuropathy, 194–196, 195f
Diadynamic currents. *See* Bernard's current
Diagnotronics, 55, 56f
Diathermy, 100
Dilatation and curettage (D&C), 61
Direct electrical nerve stimulation, 99–100
Dorsal back pain, 81
Dry needling. *See* Acupuncture

E
ECG. *See* Electrocardiogram
EEG. *See* Electroencephalogram
Elbow osteoarthritis, 82
Elbow pain, 189–190, 189f
Electric face lift, 90
Electro-sleep therapy, 98, 98f. *See also* Thermotherapy
Electroacupuncture, 165, 166f
Electrocardiogram (ECG), 3
Electroencephalogram (EEG), 3
Electrography. *See* Kirlian photography
Electromagnetic therapy, natural, 46
Electromyogram (EMG), 3
Electrophotography. *See* Kirlian photography
Electroporation, 42
Electropore, 43f
Electrotherapy, 88
  direct electrical nerve stimulation, 99–100
  high-frequency electromagnetic fields, 100–106
  TENS, 88–98

ELFMF. *See* Extremely low-frequency magnetic fields
EMG. *See* Electromyogram
Encephalins, 34
Endorphinic theory, 33
Endorphins, 34
Endoscopic retrograde cholangiopancreatography (ERCP), 64
ERCP. *See* Endoscopic retrograde cholangiopancreatography
Esophagospasm, 203, 203f
Extreme cryotherapy chambers, 79–80, 79f
Extremely low-frequency magnetic fields (ELFMF), 107, 107f–108f

F
"Failed spinal surgery syndrome", 172
Fibromyalgia, 176, 186–187
Food and Drug Administration (FDA), 91
Frostbite, 104

G
Gallbladder meridian, 132f–133f, 155b–157b
Galvanization, 88–89
Gate control theory, 33–34
Glaucoma, 210–212
Globus hystericus, 204, 204f
Governing vessel (GV), 136f–137f, 159b–160b, 172

H
Headache, 182–186
Heart meridian, 120f–121f, 146b
Heliotherapy, 3
Heliotherapy. *See* Phototherapy
Hepatic immuno-diacetic acid (HIDA), 64
High-frequency electromagnetic field, 100
  D'Arsonval's currents, 100
  microwave diathermy, 105–106, 106f
  pulsed high-frequency, 104–105, 105f
  shortwave diathermy, 101–104, 101f
Hip osteoarthritis, 81
Hip pain, 191, 192f
Huatuo points, 172
Hydro-electrical baths, 88–89
Hydrotherapy, 3–4, 80
Hyperemesis gravidarum, 209, 209f

I
IDIS. *See* Intelligent Driver Information System
Implantable pulse generator, 99f
  direct electrical nerve stimulation by, 99–100
Induction shortwave diathermy, 101–102, 102f
Infrared radiation (IR), 82–83, 83f
Insomnia, 212–213, 212f–213f
Integrative medicine, 225
Intelligent Driver Information System (IDIS), 45
Intercostal neuralgia, 175f, 176. *See also* Trigeminal neuralgia
  exemplary location of PPs/TPs, 177f

Interferential currents therapy, 90
IR. *See* Infrared radiation

J
Jump sign, 17

K
Kidney meridian, 126f–127f, 152b–153b
Kirlian photography, 15
Knee osteoarthritis, 81
Knee pain, 191–192, 193f
Krenotherapy, 6
Kyungrak System, 12

L
L1 neuralgia, 172–173, 174f
L2 neuralgia, 172, 173f
Large intestine meridian (LI meridian), 114f–115f, 140b–141b
Laser stimulation, 86–88, 87f
Lipid bilayer, 10f
Liver meridian, 134f–135f, 158b
Low back pain
  APs, 168–169
    classical, 169f
    for coccygeal pain treatment, 169f
    for foot numbness treatment, 171f
    for genital and anal regional PSSO treatment, 170f
    for sciatica treatment, 170f
  with or without sciatica, 168
Low-level laser therapy (LLLT), 87
Low-resistance point (LRP), 18
Lung meridian, 112f–113f, 140b

M
Macular degeneration, 210–212
Magnetic resonance imaging (MRI), 3, 47
Magnetotherapy, 106. *See also* Thermotherapy
  ELF-MF therapy, 107f–108f
  "soft" methods, 107
Mastitis, 104
MCT. *See* Microcurrent therapy
Medium-frequency current, 89
Meniere's disease, 200, 201f
Meridian topography, 4, 111
  conception vessel, 138f–139f
  gallbladder meridian, 132f–133f
  governing vessel, 136f–137f
  heart meridian, 120f–121f
  kidney meridian, 126f–127f
  large intestine meridian, 114f–115f
  liver meridian, 134f–135f
  lung meridian, 112f–113f
  pericardium meridian, 128f–129f
  small intestine meridian, 122f–123f
  spleen and pancreas meridian, 118f–119f
  stomach meridian, 116f–117f
  triple warmer meridian, 130f–131f
  urinary bladder meridian, 124f–125f
Microcurrent therapy (MCT), 90
Micromassage, 81
MicroTesla magnetic field, 33, 107, 108f

MicroTesla magnetic fields therapy, 108f
Microwave diathermy, 105–106, 106f
Migraine, 184
Monochrome, 86
Moxibution, 78
MRI. *See* Magnetic resonance imaging
Myofascial pain, 187

N

Nerve damage, acoustic, 199, 200f
Nervous receptors, 46
    nociceptive, 40–41
    skin, 35f
Neural convergence principle, 38f
Neuralgia, 103
Neurogenic bladder, 207–208, 207f
Neuropathy. *See* Neuralgia
Neurophysiological foundations. *See also* Reflexive physical medicine
    analgesic effectiveness, 37f
    bio-energetic therapies, 33
    bioelectrical measurements, 39f
    car manufacturers, 45
    convergence modulation theory, 40–46
    cryotherapy procedure, 38f
    damage done to internal organ, 35f
    encephalins, 34
    endorphinic theory, 33
    endorphins, 34
    gate control theory, 33–34
    history of medicine, 46
    ischaemic lower extremity, 36f
    nervous receptors, 46
    nervous system, 45
    neural convergence principle, 38f
    nociceptive signal pathway, 39f
    organ projection areas, 46–51
    physical therapies, 33
    physiological mechanisms, 34
    reflexive therapies, 33
    scientific principles, 45
    skin nervous receptors, 35f
    transmission station, 37
Neurosis. *See* Insomnia
Noctunal enuresis, 208–209, 208f

O

OED. *See* Organ electrodermal diagnostics
OPA visualization. *See* Organ projection area visualization
Open electropore, 44f
Optic nerve atrophy, 210–212, 211f
Organ electrodermal diagnostics (OED), 55
    analysis, 72
    clinical assessment
        clinical investigation procedure, 64–65
        statistical procedures, 65–66
        study design and sampling, 63–64
    clinical diagnoses and, 63–70, 67t–69t
    examination procedure, 65
    general anesthesia and surgical intervention, 61–62, 63t
    innovative aspects, 70–71
    noninvasive diagnostic method, 55
    optimal measuring parameters
        AC-based modality, 55–56
        diagnotronics, 55–56
        dry brass point electrode, 55
        electrical impedance and resistance, 55
        skin impedance, 55–56
    prospective applications for, 71–72
    rectification ratios, 59f–60f
Organ projection area visualization (OPA visualization), 15, 18
    albumin leak, 47
    contrast-enhanced MRI images, 48f, 51f
    convergence modulation theory, 46–47
    corporal OPAs, 47
    exemplary images, 48–51
    impedance and resistance measurement results, 22t
    localization
        auricular projection areas, 57–59
        bilirubin concentration and ALT activity, 61f
        physical medicine, 56
        resistance breakthrough effect, 56
    MRI sequences, 47–48
    tagged albumin, 47
    in vivo neuropeptide markers, 47
Ovarian endocrynological dysfunction, 104

P

Pain
    ankle, 193–194, 194f
    back, 172
    elbow, 189–190, 189f
    hip, 191, 192f
    knee, 191–192, 193f
    myofascial, 187
    phantom, 187
    shoulder, 188–189, 188f
    wrist, 190–191, 190f
Pain syndromes of spinal origin (PSSO), 167–168
Pericardium meridian, 128f–129f, 153b
Phantom pain, 82, 187
Pharmacotherapy, 3
Photochemical reaction, 84
Photochemotherapy ultraviolet A (PUVA), 85
Photodynamic agent, 85
Phototherapy, 82. *See also* Thermotherapy
    IR, 82–83, 83f
    laser stimulation, 86–88
    UV radiation, 83–86, 85f–86f
Physical medicine, 3
    faradic current, 3
    hot wrappings, 4
    hydrotherapy, 4
    indications, 5–6
    infrared and ultraviolet radiation, 3–4
    medical acupuncture, 5
    meridians, 4
    pharmacotherapy, 3
    physical methods, 3
    physiotherapy, 6
    piezoelectric phenomenon, 4
    reflexive therapies, 5
    rehabilitation, 6, 225
    scarifications, 4
    therapeutic effects, 3
    Western and Far Eastern methods, 5
Physiotherapy, 6
Polygraph test, 10
Post-herpetic pain. *See* Intercostal neuralgia
Pressopuncture, 109
Pressure point (PP), 34
PSSO. *See* Pain syndromes of spinal origin
Pulsed high-frequency magnetic field therapy, 104–105, 105f
Pulsed magnetic field therapy, 33
PUVA. *See* Photochemotherapy ultraviolet A

R

Raynaud's syndrome, 196, 197f
Reflexive massage, 219
    map of OPAs, 221f
    reflexology, 220
    Tsubo therapy, 220
Reflexive massage of feet/hands. *See* Reflexology
Reflexive mechanical stimulation, 109
    acupuncture, 109–214
Reflexive physical medicine, 5, 76
    acupuncture meridians, 12–13, 26–27
    acute abdomen, 75
    efficacy of, 76
    electrical impedance and resistance of OPA, 29–30
    electrotherapy, 88
        direct electrical nerve stimulation, 99–100
        high-frequency electromagnetic fields, 100–106
        TENS, 88–98
    frequency range, 76
    histomorphological structures, 30
    hormonal and humoral changes, 75
    hydrotherapy, 80
    increased metabolic rate, 75–76
    influence of organ pathology, 13–26
    internal organs and skin, 10
    magnetotherapy, 106–108
    pharmaceutical factory, 75
    phototherapy, 82
        IR, 82–83, 83f
        laser stimulation, 86–88, 87f
        UV radiation, 83–86, 85f–86f
    physiological mechanisms, 30
    reflexive mechanical stimulation, 109
        acupuncture, 109–214
    skin, anatomical structure of, 9
    skin resistance characteristics, 29
    thermographic investigation, 10–12, 28
    thermotherapy, 77–80
    transitory period of treatment, 77
    ultrasoundtherapy, 80–82
Reflexology, 220
Regional pain syndrome, 167–169, 178
    C3 neuralgia, 179–181
    cervicogenic headache, 185
    L2 neuralgia, 172
Regulatory medicine, 225

Renal colic, 75, 205–206
    APs for treatment, 206f
    softer modulation of TENS, 90
Retinopathies, 210–212
Reversible breakthrough effect, 22, 42
Rhinitis, 196–197, 198f
Ribonucleic acid (RNA), 75–76

S
Scalp acupuncture. See Craniotherapy
Scarification, 4
Selective UV-phototherapy (SUP), 85
Sensory nervous system, 225
    acupuncture therapeutic system, 30
    damage detection on body, 70–71
Shortwave diathermy, 101, 101f
    induction, 101–102
    treatment indications, 103
Shoulder osteoarthritis, 82
Shoulder pain, 188–189, 188f
SI meridian. See Small intestine meridian
Signal pathway, nociceptive, 39f
Sinusitis, 199f
    chronic, 82, 103
    oxygen in, 198
    rhinitis treatment, 199
Skin, 9, 9f
    anatomical structure, 9
        principle layers, 9
        skin appendages, 9
        stratum corneum, 9
    bioelectrical properties
        history of investigations, 13–15
        skin electrical potential investigation, 16–17
        skin electrical resistance investigation, 17–18
    electrical potential investigation, 16–18, 16f–17f
    electrical resistance, 10
    internal organs
        lipid bilayer, 10f
        reflexive therapies, 10
        skin electrical resistance, 10
    nervous receptor, 35f
    organ pathology influence, 18
        clinical investigation procedure, 20
        electrical impedance evaluation procedure, 20
        electrical resistance evaluation procedure, 20–21
        impedance evaluation results, 21–22, 21f, 23f
        OPA, 18, 19f
        resistance evaluation results, 22–24, 23f–24f
        statistical procedures, 21
        study design and sampling, 19
    thermographic investigation, 11f–12f
        clinical diagnoses, 11
        insertion of needle, 12
        temperatures, 10
        thermal effect, 11
        trigger points, 11–12
Slimming therapy, 90. See also Thermotherapy
Small intestine meridian (SI meridian), 122f–123f, 146b–147b
Soft X-rays, 83–84
Solar radiation, 82
    IR radiation, 82–83
    laser stimulation, 86–88
    UV radiation, 83–86
SP meridian. See Spleen pancreas meridian
Spleen pancreas meridian (SP meridian), 118f–119f, 144b–145b
Stationary induction electrodes, 101–102
Stomach meridian, 116f–117f, 142b–144b
Stratum corneum, 9
SUP. See Selective UV-phototherapy

T
T11–12 neuralgia, 175–176, 175f
Tagged albumin, 47
Tendonitis, 82, 103
Tennis elbow, 189
TENS. See Transcutaneous electrical nerve stimulation
Thermography, 10
    ischaemic lower extremity, 36f
    skin investigation, 10–12
Thermotherapy, 77
    therapeutic application
        of cold, 78–80
        of heat, 77–78
    extreme cryotherapy, 79, 79f
Tinnitus. See Nerve damage, acoustic
TP. See Trigger point
Traebert's current, 88–89
Transcutaneous electrical nerve stimulation (TENS), 5, 10, 88–89
    electro-sleep therapy, 98
    electrodes placement, 92f–97f
    electrotherapeutic parameters, 89, 89f
    indications for, 91
    interferential currents therapy, 90
    MCT, 90
    "softer" modulation, 90
    treatment with double-channel device, 91f
Transitory period of treatment, 77
Trigeminal neuralgia, 82, 92f–97f, 181, 183f
    mandibular division treatment, 182f
    maxillary division treatment, 182f
    ophthalmic division treatment, 183f
Trigger point (TP), 11–12, 81
Triple warmer meridian, 130f–131f, 154b–155b
Tsubo therapy, 220

U
Ulcerative colitis, 204–205, 205f
Ultra Reiz. See Traebert's current
Ultrasoundtherapy
    mechanical vibration, 80–81
    micromassage, 81
    by mobile applicator, 81f
    parameters, 81
    physical medicine, 4
    "Western" physical therapy, 33
Ultraviolet radiation (UV radiation), 3–4, 82, 85f–86f
    causing photochemical reaction, 84
    electromagnetic radiation, 83–84
    photochemical erythema evolution, 84
    prophylactic application, 85
Unconventional medicine, 225
Upper back pain, 178f
    C3 neuralgia, 179–181
    cervicogenic headache, 185
    electrostimulation program, 90
    with or without brachialgia, 177–178, 187
Urinary bladder
    meridian, 124f–125f, 148b–151b
    urine for microscopy, 20, 20
UV radiation. See Ultraviolet radiation

V
Visual aura, 167–168, 184

W
Whip shower, 80
Wrist pain, 190–191, 190f
    physical medicine use, 191

Z
Zu-sanli point, 12–13

Printed in the United States
By Bookmasters